HEIDELBERG SCIENCE LIBRARY ▍Volume 20

The Invisible Universe

The Story of Radio Astronomy

Gerrit L. Verschuur

The English Universities Press Ltd. London
Springer-Verlag New York Heidelberg Berlin

1974

Gerrit L. Verschuur
Fiske Planetarium
Department of Astrogeophysics
University of Colorado
Boulder, Colorado 80302

Library of Congress Cataloging in Publication Data
Verschuur, G L 1937–
The invisible universe.
1. Radio astronomy. I. Title.
QB475.V47 522′.682 73-22202
ISBN 0-387-90078-0

Printed in the United States of America.

ISBN 0-340-19180-5 The English Universities Press Ltd. London
ISBN 0-387-90078-0 Springer-Verlag New York Heidelberg Berlin
ISBN 0-540-90078-0 Springer-Verlag Berlin Heidelberg New York

PREFACE

Astronomy is one of the oldest sciences and at the same time one of the most popular of the modern sciences. It is well within the grasp of most people to experience the enormity of the universe by gazing at the stars at night and beholding the majestic sweep of the Milky Way across the otherwise pitch black void of space. Man can let his imagination wander outward and let it travel through the depths of space past star fields, through clouds of dust and gas, and outward beyond the limits of our galaxy. By using a telescope he can see what it looks like out there, where countless galaxies drift through space, and he can see nearby exploding stars and incandescent gas clouds in our own Milky Way. All these things are part of the universe that our eyes can see.

But there exists another universe out there, one in which many objects can never be seen, even with the most powerful optical telescopes. It is the universe from which radio waves carry information about the existence of phenomena never previously dreamed of by man. The science of radio astronomy has brought to man knowledge of strange new objects such as quasars and pulsars and knowledge of the existence of many complex molecules existing in clouds drifting between the stars—molecules essential to the beginnings of life on our planet and probably on other planets as well.

Giant dish-shaped radio telescopes and complex radio receivers are the tools of this new science, which was born in only the early 1930s as the result of an accidental discovery by an engineer named Karl Jansky.

The radio sky is quite different from the one we are used to seeing at night, but is just as important to our understanding of the universe around us. This radio sky contains thousands of radio sources, some of which can be seen optically, but many of which are still invisible to earth-based optical telescopes. It is a sky that shines brightly at some wavelengths and not at others, so that as the radio astronomer tunes his radio set across the radio band the whole sky takes on a different appearance. Radio astronomy is also a science that can be pursued by day as well as by night, since the radio waves are always "visible," regardless of the presence of sunlight or clouds. These radio waves can also penetrate the thick clouds of obscuring matter that exist in the Milky Way, clouds that prevent the optical astronomer from seeing very far into our own galaxy.

Our tale is about a science born in the mid-twentieth century and now beginning to flourish with dramatic discoveries of objects and phenomena completely undreamt of even 10 years ago. Who would have suspected that there are objects in the universe strongly pulsating at incredibly regular intervals—so regular, in fact, that only the most accurate clocks on Earth can measure their rates. The pulsing signals from the pulsars will ultimately allow man to make some remarkable measurements, such as the shape of the Earth and the distance between continents, with incredible accuracy.

Ten years ago no one would have predicted that radio astronomers would be routinely discovering an ever-increasing number of complex organic molecules in clouds existing in the so-called empty reaches of space between the stars, molecules that on Earth were precursors to the development of life. These molecules coexist in interstellar clouds from which stars and planetary systems ultimately form, and therefore astronomers are now beginning to seriously consider that life may be very widespread in the universe.

In 1960, quasars were unheard of, but now these remarkable objects are being routinely studied and their name is common knowledge outside the astronomical community. Are they distant or nearby objects? Are they the birth or the death throes of galaxies? Where do the enormous energies come from that are clearly required to explain their emissions?

In 1960, no one realized that the entire universe was bathed in a soft glow of radiation now thought to be the indications of the remains of the possible "Big Bang" that might have started the whole universe evolving.

In this book we explain, in the simplest possible terms, the hows, whys, and whats of this young science, which is expanding so fast. In the past it has been usual to include a brief mention of radio astronomy as a footnote, or at most a chapter or two, in basic books on optical astronomy—books that have usually been written by optical astronomers. Although the two branches of astronomy are closely related, we shall discuss astronomy from the radio astronomer's point of view because the discoveries revealed by radio can be appreciated without a knowledge of all aspects of optical astronomy. An elementary knowledge of some of the broader concepts of astronomy in general will no doubt help in the appreciation of the discoveries made by radio astronomers, but since the techniques used by the radio astronomer are so different from those usually pictured in astronomy and because the phenomena studied are often quite invisible to the eye and the telescope, we must approach the subject on its own terms.

We should bear in mind that, based simply on the staggering number of discoveries in the last decade, future discoveries might radically alter our picture of the universe even more in the next decade. Many astronomers feel that the probability of our detecting the presence of intelligent life elsewhere in the universe in the not too distant future is fairly good and that the day when we can prove, without doubt, that life must be common in the universe is drawing ever nearer. The man in the street often asks why we are interested in astronomy, and in particular in radio astronomy, since it is apparently far removed from man's direct sensory experiences and has little obvious usefulness. While we can all see the stars and galaxies with even a small telescope, and can experience their existence directly, we have no sense organ to detect radio signals. So why bother with the subject at all? Man has always tried to answer the question: "Where do I come from?" The study of astronomy has often been directed to answering this question, concerning the origin of the stars, the planets, and the universe as a whole. Radio astronomy is essential to the quest for such answers, since some of the most basic materials of the universe, such as hydrogen gas and organic molecules, are observable only by means of radio astronomy. Objects such as quasars and pulsars are also best studied by looking at the radio waves they transmit. All these objects fascinate explorers in astronomy, and their study leads to a better overall knowledge of the universe around us.

Also, other civilizations are very likely to be signaling between themselves by means of radio waves, and what more dramatic discovery in science could there be than finding life elsewhere! Such a discovery might well have far-reaching effects to life on this planet because man's earth-centered view of the universe will forever be destroyed and he will become, hopefully, more aware that he is merely traveling through space on one of many millions of spaceships with limited supplies of natural resources and food.

Twice already, in the 1960s, discoveries by radio astronomers have been made that were so surprising that for some time an artificial origin for the signals was seriously considered by many. The best known example is the discovery of the "pulsars" at Cambridge in England, a discovery that was kept a closely guarded secret for several months to prevent the media from publicizing the first explanation suggested—that the signals were indicators of other civilizations.

During our story we shall also see how important luck can be in making discoveries in this fast-moving new science. Several new phenomena were discovered, merely by chance, during experiments designed for quite different purposes. Who knows what surprises the future holds in store for us.

We shall also discuss the actual details of some of the "rat races" that take place in such a fast-moving research field. Many people have the image of the scientist as a careful, relatively unemotional person in a white coat, dedicatedly following a laborious piece of work through to its conclusion for relatively little reward. The truth is often the opposite extreme. Many of the best research workers are temperamental and similar to artists in that they are usually very emotionally involved in their work, and work best when they are inspired, for whatever reason, but might be unable to do anything useful when that undefinable something is missing. Personal ambition plays a very important role. In the present system, the number of papers published is of the greatest importance for personal advancement, and the way to have papers published is to make new discoveries. Big discoveries hopefully bring correspondingly bigger rewards. One therefore finds the bandwagon effect playing a role; that is, a scientist may suddenly redirect his research interests into a fast-opening new branch of research in the hope of making new discoveries in a way somewhat akin to a prospector searching for gold in a gold rush. This appears to be largely a phenomenon of the U.S.A.! For example, the discovery of pulsars set off a frantic rush in many observatories to find new pulsars, which were then named after the observatory concerned. The discovery of the first three-atom molecules in 1969 also set off an unbelievable rush to observe new molecules, a bandwagon that is still rolling. One can liken the search for new molecules, or pulsars, using radio telescopes, to a gold rush where every man is for himself and where teams form only when individuals have to band together for survival in the wilderness of competition. The only limiting factor in such rushes at present is the lack of sufficient radio telescopes.

Because radio astronomy is so young and vital, it is still possible for the newly entered graduate student to make important discoveries, even if only as a result of a lucky break. Such events will become increasingly rare as the science matures, and one therefore does not find this quite so often in, say, optical astronomy, where many of the important breakthroughs in recent years have come as a result of much painstaking work by a person, or persons, of established reputation who had the necessary experience to make the important step forward.

The present state of radio astronomy is much like the time during which Galileo was first pointing his telescope at the stars. It was difficult for him not to make many new discoveries. So it is in radio astronomy today. Almost whatever is done, or wherever the telescope is pointed, new things are found, and the number of problems now needing solutions keeps growing.

A final point to help view the present situation in radio astronomy in the context of history is to note that it was only in the first third of this century that man realized that the Milky Way was but one of many galaxies in the universe, that the space between the stars was far from empty, and that some idea was obtained of just how big the universe might be. Only then were the first estimates of the age of the universe made, as well as just how far away and how old the stars in fact are. Bear in mind that radio astronomy started to grow only slowly after World War II and that the presence of hydrogen gas between the stars was revealed by radio means in only 1951. Studies of this gas have since given us our first impression of what the Milky Way Galaxy might look like from the outside. It was only in 1963 that the quasars were discovered, in 1967 that pulsars were found, and in 1968 that the first triatomic molecules in space were found. The first diatomic molecule was found in 1963, but since the beginning of 1970 the presence of no fewer than 30 complex molecules in space has been revealed. In the early 1960s, only 500 or so radio sources were known to exist. Now some 40,000 are catalogued. Many of these are invisible optically for no known reason.

Radio telescopes capable of "photographing" radio objects in the universe, nearly as clearly as optical telescopes can, have just been brought into operation and others are planned. Who knows what incredible things they will reveal? Join us in our journey through the radio astronomer's world as we examine how he sets about his work and what he has discovered. We shall relate the story of radio astronomy as it is today in a completely nonmathematical way intended for the amateur astronomer, the planetarium goer, the college student taking an introductory astronomy course, or the high school teacher or student wishing to know more about the science. As the number of startling discoveries increases, everyone should be able to obtain a basic idea of what the radio astronomer does and how the taxpayer's money is being used to help his research.

I wish to acknowledge the help of Donna Beemer, who typed the manuscript, and Jon Spargo, who read an earlier version. I am also particularly grateful to those colleagues from all over the world who sent photographic material for use in the book and to the NRAO for making so many photographs available from their files.

G. L. Verschuur

CONTENTS

Preface v

Chapter 1. Radio Signals from Space and How the Radio Astronomer Picks Them Up 1

- A Definition of Radio Astronomy 1
- The Difference Between Light and Radio Waves 1
- Radio Telescopes 3
- Visual and Radio Astronomy 6
- How Radio Astronomers Detect Radio Waves from Space 7
- What Can Be Measured 9
  - Mapping 11
  - The Spectrum 11
  - Polarization 13
  - Variability 13
- Why Radio Telescopes Are Not Shiny 14
- Other Waves in the Electromagnetic Spectrum 16
- Windows into Space 16
- Direction Finding in the Radio Wave Region 17
- Protected Bands in the Radio Wave Region 19

Chapter 2. The Birth of Radio Astronomy 21

- Tracking Interference 21
- Radar Jamming by the Sun 24
- Radio Signals from Gas Clouds 25
- Early Days at Jodrell Bank 26
- The U.S. National Radio Astronomy Observatory 29

Chapter 3. The Radio Sun and Planets 32

- Radio Signals from the Sun 32
- Solar Storms and Explosions 34

Storm Bursts 36
"Photographing" Solar Radio Bursts 36
The Planets 39
Black–body Emission 39
Jupiter 40
Jupiter's Bursts 41
The Greenhouse on Venus 42
The Other Planets 42
Radar Astronomy 44

Chapter 4. Producing Radio Signals in Space 47

Nonthermal or Synchrotron Emission 47
Thermal Emission 47
Plasma Oscillations 48
Spectral Line Emission 49

Chapter 5. The Milky Way Radio Beacon 50

The Milky Way: Our Galaxy 50
Radio Signals from the Milky Way 50
Mapping the Milky Way 51
What Are Spurs? 52

Chapter 6. Exploding Stars 54

Guest Stars Make Their Appearance 54

Chapter 7. Interstellar Space: Is It Empty? 58

Radio Signals from the Building of the Universe 58
Taking the Temperature of Hydrogen Clouds 61
The Doppler Effect 61
Looking at the Milky Way from the "Outside" Using the Hydrogen Line 63
High-Velocity Clouds 64

Chapter 8. Polarization in Astronomy 68

Magnetic Fields in Space and Polarization of Radio Waves 68
The Polarization of Starlight 71
Faraday Rotation 72

Chapter 9. Molecules Between the Stars 74

Line Emission from Atoms and Molcecules 74
Laboratory Checks 75
The Story of OH and Mysterium
Maser Amplification in Space 77
Water, Ammonia, and Embalming Fluid 78
Other Interstellar Pollutants 81
The Bandwagon Effect 83
Astrochemistry 83

Chapter 10. Locating Radio Sources in the Universe 85

What Is a Radio Source? 85
Locating the Position of Radio Sources 85
The Moon as a Shield 89
The Identification of Cygnus A 89
How Far Are Radio Sources—The Redshift 93

Chapter 11. Exploding Galaxies 94

Double Radio Sources 94
Jets of Matter 94
Explosion in Galaxies 97

Chapter 12. The Enigma of the Quasars 98

The Story of Quasar 3C273 98
A Radio "Star" Is Pinpointed 100
Objects at the Edge of the Universe 101
Variability on the Cosmic Scale 103

Long Baseline Interferometers 104
Intercontinental Interferometers 106
New Ideas 109

Chapter 13. A Journey to a Quasar 111

Chapter 14. Little Green Men, White Dwarfs, or Neutron Stars? The Story of Pulsars 115

An Accidental Discovery 115
Some Properties of Pulsars 118
The Game of the Names 121
Discovery of the Crab Pulsar 121
A Blinking Star 123
What Is a Pulsar? 125
Star Quakes 126
Pulsars as Probes of the Interstellar Medium 126
Pulsars and Their Motion Through Space 128

Chapter 15. Radio Stars 129

Cygnus X3. What Is It? 132

Chapter 16. Radio Cosmology 135

Is the Universe Expanding? 135
Other Causes for a Redshift 137
Time Variations of Physical Constants 138
The Consequences of a Nonexpanding Universe 139
Three Degrees Everywhere 140
Testing Einstein's Theories 141

Chapter 17. Radio Telescopes of the World 143

The World Distribution of Radio Telescopes 143
Some Large Interferometers 148

Aperture Synthesis 149
Radio Telescopes of the Future 151
Project Cyclops 155

Chapter 18. How a Radio Astronomer Makes His Observations and Studies the Data 157

Preparation 157
Observation 158
Publication 159

Chapter 19. Is Anyone Else Out There? 161

Life in the Universe 161
Interstellar Molecules and the Formation of Planets 161
Searching for Extraterrestrial Life 164
Should We Transmit Instead? 167
Can We Receive Their Signals Now? 167

Index 169

The Invisible Universe

The Story of Radio Astronomy

CHAPTER 1

RADIO SIGNALS FROM SPACE AND HOW THE RADIO ASTRONOMER PICKS THEM UP

A Definition of Radio Astronomy

Stars, galaxies, the Sun, the planets, and many other objects radiate not only light waves, but also strong radio signals. Some objects transmit only radio waves and will forever be invisible to the human eye. These radio waves fill the air around us and can be extracted with suitable antennas (Figure 1) and radio sets tuned to the right wavelengths. Radio astronomy is the science that studies the radio waves reaching Earth from space. The radio astronomer does not transmit radio signals; he only tunes in to those coming to him from beyond Earth's atmosphere.

The Difference Between Light and Radio Waves

Both light and radio waves are aspects of the same phenomenon; only their wavelengths are different. They are both electromagnetic radiations, an impressive word that merely describes their physical nature. Light waves are short wavelength radiations whereas radio waves are much longer. Let me explain a few of these terms.

The word spectrum refers to the whole range of wavelengths we are considering, but what does wavelength mean? It refers to the distance between crests in a wave pattern such as might be produced when a stone is thrown into a clear pond. The source of the wave pattern in the pond is the disturbance produced by the stone on striking the water. In exactly the same way, radio and light waves are caused to radiate from a star as a result of some type of disturbance produced in its surface.

Fig. 1—The site of the National Radio Astronomy Observatory at Green Bank, West Virginia. On the right are the three 85-foot-diameter radio telescopes that make up the interferometer. In the left foreground is the 300-foot-diameter telescope, and behind it the 140-foot telescope. These telescopes study radio signals from space 24 hours a day. (Courtesy NRAO.)

Electromagnetic waves also have a wave pattern, which is roughly related to the strength of the radiation at any point. This radiation is successively greater and smaller, or, if one could look at the wave pattern radiating from the star at any instant, one would find that the radiation was alternately stronger and weaker as one moved out from the star. The crests and dips in the surface of the water on the pond are therefore replaced by increases and decreases in the strength of the light or radio signals being sent out. One can measure the wavelength of the radiation by finding the distance between successive peaks in the wave pattern.

The wavelength of the light to which the human eye is sensitive is around 5000 Å (angstroms), which is equivalent to 50-millionths of a centimeter. Radio waves, however, range in length from a few millimeters to hundreds of meters, the later being the wavelengths we receive with our broadcast band radios.

Fig. 2—The 300-foot-diameter radio telescope of the National Radio Astronomy Observatory at Green Bank, West Virginia. This telescope was used to discover the pulsar in the Crab Nebula. Radio signals bounce off the metal-mesh surface to gather at the focus point (the top of the two masts above the surface). An antenna placed at the focus collects the signals and they are sent by cable to the receivers. (Courtesy NRAO.)

Radio Telescopes

The air around us is filled with many invisible radiations. The ones with which we are most familiar are the radio waves transmitted by local and distant radio stations. We can extract these signals from the surrounding air by using a radio set tuned to the right wavelength. Besides these signals, there are also radio waves which have traveled from the furthest depths of space to reach Earth. Unfortunately our short-wave radios or our TV sets are not sensitive enough to pick up these transmissions from space; so radio astronomers have to use giant antennas (Figure 2) and very sophisticated radio sets to tune in to the cosmic radio stations. The signals they pick up reveal the presence of a whole range of unexpected phenomena in the universe. They find some types of objects never seen before and some objects that will forever remain unseen because they do not appear to send out any light waves.

We are all familiar with the fact that stars give out light that we can sense with our eyes when looking at the sky on a clear night. If we want to see stars farther away or fainter than those visible to the unaided eye we have to use a telescope. This is a device, often consisting of a curved mirror, that collects all the light reaching an area much larger than that of the human eye and reflects the light to a single point called the focus. By placing his eye or a camera at the focus, the astronomer can see many more stars than would ordinarily be visible to him, and the larger the mirror he uses, the fainter and the farther away are the objects that he is able to study. Telescopes that work by the principle of reflecting large quantities of light in this way are called *reflecting telescopes*.

Besides light, there are other types of radiation, such as radio waves, reaching us from outer space. Since man can neither see, feel, nor hear radio waves, he needs the aid of radio sets to convert the signals into some form that he can sense, for example, a visible record or a sound on a loudspeaker. He therefore needs to construct giant antennas and connect them to radio sets sensitive enough to pick up the incredibly weak signals from outer space. That is what a radio telescope helps him to do.

We are all familiar with the appearance of TV antennas mounted on rooftops and know that in order to pick up more and more distant TV stations one has to use bigger and bigger antennas, or simply more of the same type connected so that more signals can be collected. Since the stars are very far away indeed, one requires enormous antennas to pick up the radio signals they transmit. In fact one would need to cover many acres of ground with TV type antennas in order to pick up the radio signals from most astronomical bodies. Some radio observatories have done, and some still do, just this (see Figure 3). Notice that when we use the term *weak* signal we are comparing it to the signals we would normally pick up on our TV or radio antennas at home, when tuning into a distant station. Although the cosmic radio transmitters, such as quasars, are inherently tremendously powerful transmitters, the signals are enormously diluted in their journey through space to us and we therefore only receive extremely weak signals.

There is an alternative to constructing enormous arrays, as they are called, of small antennas covering acres of ground, and that is to build a large reflecting telescope. Radio waves are known to bounce off metal surfaces and therefore, if one constructs a curved (parabolic) metal surface similar to, but much larger than, an optical mirror, all the radio waves that reach its surface from a particular direction will be reflected to a single point also called the focus. Now we need only place a single, small antenna at the focus in order to intercept all the radio waves that might otherwise have been received over an area of ground equal to that of the dish itself.

Such a metal reflector is called a radio telescope and is exactly similar in principle to the optical reflecting telescope. The main difference is usually

Fig. 3—Part of the four acres of radio telescope of Cambridge University, which first found the radio signals from pulsars. (Courtesy Mullard Radio Astronomy Observatory.)

one of size and the fact that the radio telescope surface is made of metal whereas the optical telescope is made of silvered glass. A shiny mirror reflects light waves but a metal mirror, whether shiny or not, is needed to reflect radio waves. We shall discuss later how the reflecting properties are determined by the smoothness, or shininess, of the metal surface.

In order to suspend the antenna at the precise focus, a support structure, often a tripod or quadropod, of steel girders is used, and the antenna is held atop these support legs.

The largest movable radio telescopes in the world are over 100 yards in diameter (that is, as large as a football field), and yet these gigantic steel structures can be directed with incredible precision to any point in the sky. Thousands of tons of steel can be made to move on enormous bearings, guided by computers that respond to the instructions fed to them by radio astronomers and specially trained telescope operators.

The term "radio telescope" is usually thought of as including the reflecting bowl as well as the antenna mounted at the focus. An alternative word, used by many astronomers to describe a radio telescope is a "dish," referring, of course, to the shape of the giant reflectors. Radio telescopes do not always have this characteristic shape, however, since the description of the radio telescope is equally well applied to any system of antennas used to collect radio waves from outer space. For example, a field full of small antennas or groups of many dishes could equally well be called a radio telescope.

Visual and Radio Astronomy

Broadcast signals are normally thought of in terms of the sounds they produce in a radio set, and since we have already given an outline of the radio telescope as being a somewhat sophisticated space radio set, it would be very easy to form a picture of a radio astronomer listening in on the universe. Indeed the radio astronomer can listen to the signals he receives if he wants to, and some of the sounds produced by distant objects in the universe make very exciting listening, such as those produced by the pulsars. However, this is no more his main preoccupation than is an optical astronomer's to gaze perpetually through his telescope. In fact for the most part, research astronomers spend relatively little time actually looking or listening, but mostly in making permanent records of what they observe; they study these records later, away from the telescope.

The optical astronomer does this by using light-sensitive photographic film and a camera and the radio astronomer by using sophisticated tape, or paper chart, recording equipment, which will give him a record of the way the signals varied during his observations. These can then be more closely studied later and further processed in a computer if necessary. The radio astronomer may also look at an immediate display of the signals he receives on a pen recorder or on a screen similar to a TV screen. In many experiments he can therefore tell immediately if his research is producing any results, whereas during other experiments he might have to collect information for many days at a time and afterward examine it in a computer before he knows what the work has produced. The reasons for this are discussed later in this book.

So the story we shall relate is very different from the one told about the work done with the largest optical telescopes on earth, such as the 200-inch Mount Palomar telescope. We are dealing here with a world of antennas, enormous metal dish-shaped reflectors, radio sets, pen recorders, tape recorders, and computers, and not a world of photographic plates, highly polished mirrors, and how many hours of good seeing are available at any particular mountain top, since radio astronomy is generally an all-weather science and can be undertaken day and night.

Some readers might not be particularly interested in knowing, at this stage, the details of how radio astronomers go about their business of studying radio signals from outer space, and they may instead turn either to page 032 to find out about the historical beginnings of radio astronomy or to Chapter 3 where we relate the types of phenomena or objects that have recently been discovered in space. These readers might return to this chapter at a later point in order to increase their understanding, and hopefully their enjoyment, of this book.

How Radio Astronomers Detect Radio Waves from Space

The radio astronomer does not listen to the signals produced by his radio set in a loudspeaker, but instead looks at them in the sense that he observes their intensity and the variation of this intensity on a meter of some sort. This meter can also be connected to other devices such as pen recorders or tape recorders to allow a permanent record to be made. Let us pause briefly and explain some of these terms and concepts.

As mentioned above, the dish-shaped radio telescope reflects radio waves striking its surface to a single point at which the antenna is placed. An antenna is a device through which electrical current flows in response to the radio signals that strike it. These currents are then sent down an electrical cable that threads a path down the telescope to the laboratory in which the radio astronomer has placed his radio set (see Figure 4). Such a radio set is usually referred to as a radio receiver or amplifier, which amplifies or increases the strength of the signals received so that they become detectable. It is capable of translating radio signals into some comprehensible form. In our home radio sets this form is a series of sounds on a loudspeaker, usually spoken words or music. However, the sounds produced by the cosmical radio

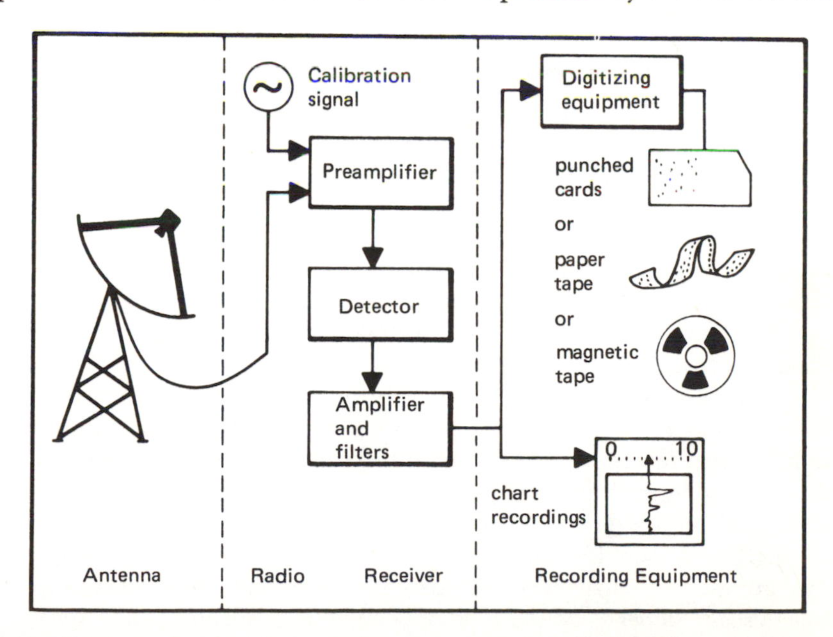

Fig. 4—Essential elements in a radio-astronomical system. (Courtesy NRAO.)

stations are not easy to hear on a loudspeaker since they are usually just too weak for the human ear to pick out from the other sounds produced by the receiver itself.

This brings us to the first point that illustrates what type of signals the radio astronomer is picking up. Many readers may have noticed that there is a steady hissing produced by their FM receivers or their TV sets if these are tuned away from some transmission. This sound is called *noise* and is produced within the radio set itself.

The radio astronomer always has this noise signal in his receiver, and any additional signal he picks up within the radio telescope has to be picked out from this intrinsic receiver noise. Unfortunately the cosmic radio transmitters generate signals that are identical in character to receiver noise, in particular in the sounds they generate. In fact, most of the radio signals produced in cosmic bodies are also produced by the motions of electrons at very high temperatures—much higher than those in the vacuum tubes.

If a radio telescope is picking up radio waves from a particular object in space, these will only manifest themselves as a change in the strength, or loudness, of the noise in the receiver. If a loudspeaker is used, the increase in this loudness is so small for the large majority of celestial radio transmitters that the human ear is just not sensitive enough to pick it up. Exceptions to this are the very strong emitters of radio waves, such as the Sun and Jupiter and one or two supernovae located relatively near to the Sun, which produce very strong signals, even in simple radio telescope systems. Instead, the radio astronomer watches the deflection of a meter that is connected to his receiver. Many FM receivers have meters of this type, which allow a person to tune in accurately to weak stations. In the same way, the radio astronomer can detect the presence of a new radio *source*, as they are usually called, by watching for small deflections on the meter while the radio telescope scans the sky in search of new objects.

To obtain a permanent record of the observations being made, the radio astronomer could write down, on some convenient note paper, the levels indicated by the meter every 10 seconds or so, but that would clearly become a very tedious business. To avoid having to do this he incorporates an automatic recording device into his receiver. These devices range from paper chart recorders, which feed out paper at a constant speed and have a pen driven by a motor to draw a line according to the signal being received, or magnetic tape recorders, which record a set of numbers in a form that can subsequently be read by a computer. To do this the receiver has to feed the radio signals into a device that converts the electrical signals into numbers, and such a device is called an *analog-to-digital converter.* Many radio observatories use paper tape or cards for collecting the information; the choice usually depends on the type of computers available.

Nowadays the pen recorder is hardly ever used as a data recording device, although it was still common in the early sixties. Instead it is used as a monitor of the way the receiver is acting and can quickly show the astronomer if any interference is being picked up or if the receiver has failed in some way.

Examination of the way a pen recorder responds to the presence of a source of radio waves in the direction in which the telescope is pointed may be useful at this stage. First of all, if no signal from space is being received, the pen will draw a wiggly, but nearly smooth line along the paper as it feeds out of the machine. The wiggly nature of the line is produced by small random variations of the noise in the system. This zero level is referred to as the *baseline*. When a signal is received, the pen deflects by an amount that depends on the strength, or loudness, of the received radio signals. Since noise is always present in the system, even the line from a strong signal will have some wiggle. One exception to this is that pulsars (see later), which produce short intense signals, and the pen will move rapidly back and forth across the paper. As the telescope moves away from the position of the radio source, the pen will return to its original level and remain there until more radio signals are detected. This is illustrated in Figure 5.

If you want to hear what the radio astronomer would in driving his telescope past the position of a very strong radio object, just switch on an FM set, turn it to an unused part of the dial, slowly turn up the volume control, and then turn it down again.

By marking the chart recorder with regular time markers, the radio astronomer has an accurate record of when the deflections occurred, and since he also keeps a record of where the telescope was pointing at any particular time he has the information necessary to allocate a position in the sky for the source of the radio waves. This allows him to look at a photograph of that part of the sky in order to see if he can associate any visible object with his newly discovered radio source.

If the data being collected is put onto a magnetic tape, then the positional information can obviously be recorded too. Sometimes a small computer is available at the telescope which will record and process the information being received and produce only the information the astronomer desires. These *on-line computers*, as they are called, make the data collecting and interpretation much easier for the radio astronomer. If the astronomer does produce a tape or cards containing his observations, it or they have to be processed by an *off-line computer*, usually located well away from the telescope.

What Can Be Measured

There are only a limited number of well-defined types of measurements a radio astronomer can make. Clearly he is confined to examining only those

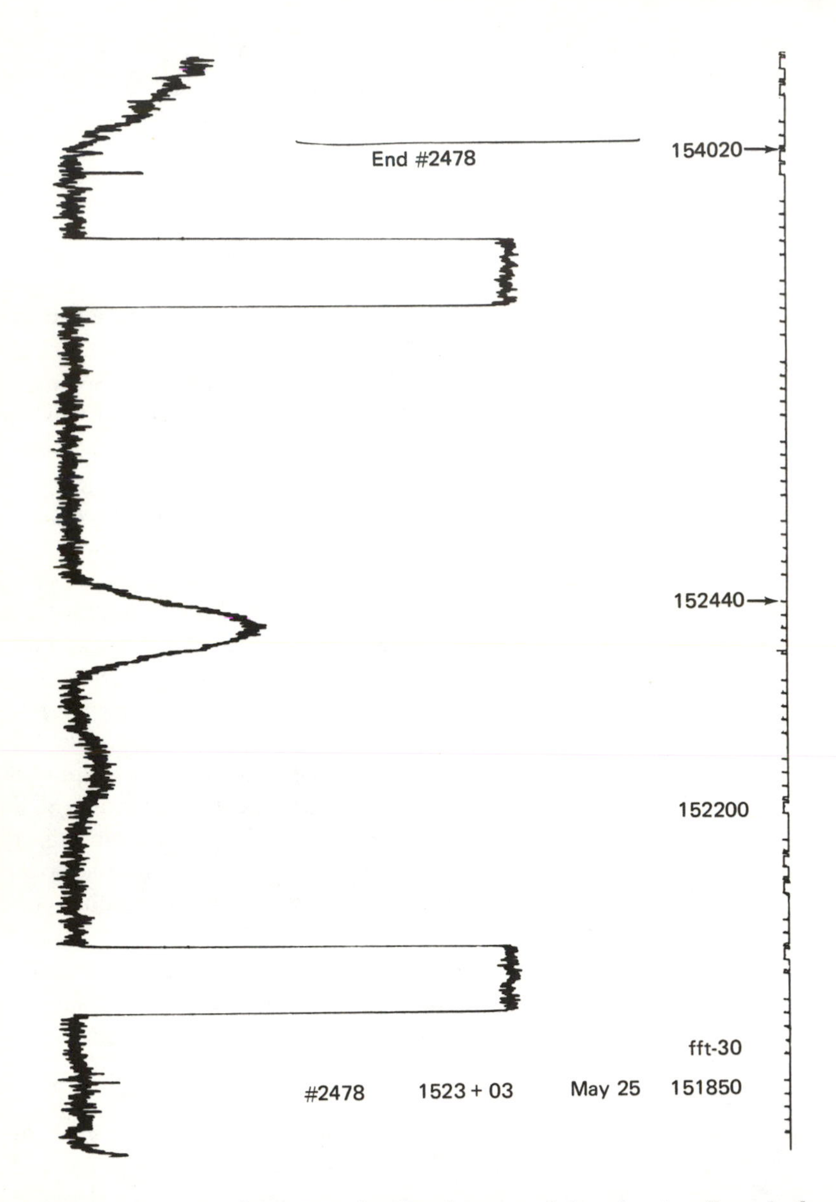

Fig. 5—A chart recorder output showing the type of data that is collected when a radio telescope is scanned across the position of a radio source seen as a deflection in the middle of the scan. Noise is visible in this scan, and the two large vertical deflections are produced by the calibration signal that is activated by the radio astronomer so that he can calculate the intensity produced in his receiver by the radio source. This radio source might have been an unequal double or it might be two separate sources. The radio astronomer would have to examine photographs of the region of the sky he was scanning in order to identify the origin of these radio signals. (Courtesy NRAO.)

radio waves that actually reach the surface of the Earth, which means that he is only able to observe radio waves shorter than about 60 meters long, because at longer wavelengths radio signals do not penetrate the highest reaches of the Earth's atmosphere, and longer than a few millimeters, because anything shorter is absorbed by the atmosphere. Let us first list those things that can be measured and then elaborate a little in order to clarify the concepts. The radio astronomer can measure:

(a) the variation of signal received with position in the sky,
(b) the variation of the signal as the radio set is tuned to different wavelengths across the radio band,
(c) the polarization, and
(d) the variation of the signal with time.*

Mapping

The first of these is a measurement of how strong the radio signals are as the telescope is moved. The object giving rise to the radio waves will have some definite position in the sky and might well have a structure that needs to be determined. For example, it might be a round object, or sausage-shaped, or ring-shaped, and it might be located at the position of a well-known optical object such as a star or a galaxy (see example in Figure 6). The radio astronomer will present his results as a *map* of the radio object using, as his scale, the two position indicators (coordinates) commonly used in astronomy —right ascension and declination. When these two position coordinates have been determined, any other astronomer can subsequently point his telescope to the same spot and expect to "see" the same object.

The Spectrum

While a given position in the sky is being tracked, the variation of the received intensity as the wavelength of the receiver is changed gives important data about the radio waves received. In the same way that stars have a very definite spectrum (in other words, they might have relatively more red or blue light in them), so the radio source has a very particular spectrum. For example, it might be brighter at shorter or longer wavelengths or it might have the same brightness at all wavelengths. In any case, the

* *Antenna temperature*—The units used by radio astronomers to indicate the intensity of the radio signals being received by the radio telescope. The antenna indirectly measures the temperature of distant objects in space. Temperatures are measured in units of degrees absolute (or Kelvin) above absolute zero (0°C = 273°K). The deflections on the chart recorders connected to the radio receivers can be calibrated by injecting a radio signal of known temperature into the antenna or by pointing the radio telescope at a radio source in the sky whose temperature has been calibrated already.

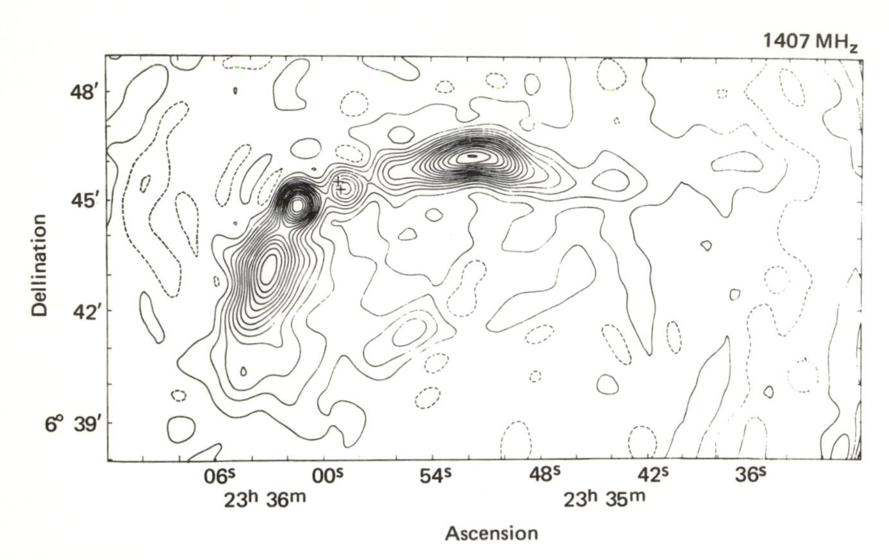

Fig. 6—A radio contour map of the radio source 3C465 made with the Cambridge radio telescope at a frequency of 1407 MHz (wavelength 21 centimeters). The two crosses indicate the position of the double radio galaxy NGC 7720. (Courtesy Mullard Radio Astronomy Observatory.)

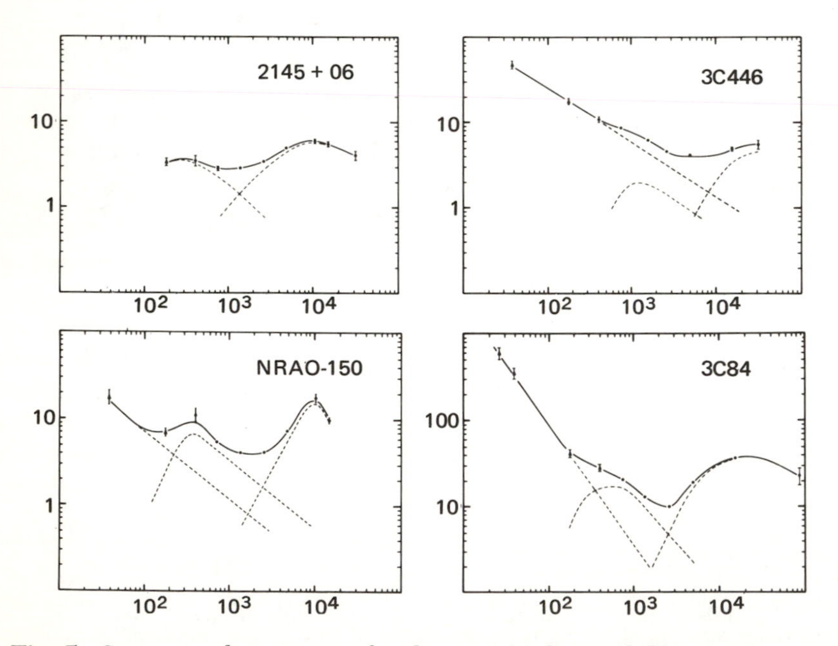

Fig. 7—Some complex spectra of radio sources that probably contain several centers of emission, possibly as the result of successive explosive outbursts. The intensity of the received radio radiation (in flux units) is plotted as a function of frequency (in MHz). (Courtesy NRAO.) (See table 2 for definition of flux units)

strength of the radio waves emitted by the source will depend on wavelength in some way that has to be determined for each source, because it turns out that very few radio sources have similar spectra. A knowledge of the spectrum allows the radio astronomer to describe more accurately the physical conditions existing within the radio object iself, whether that object is a quasar, a galaxy, or a star. There are a few basic types of spectra that the radio astronomer finds, but in general he cannot, as a result of a single measurement or even the measurement at two wavelengths, predict what the rest of the spectrum of the radio source would look like. Examples of such spectra are shown in Figure 7.

Polarization

The third parameter an astronomer needs to measure is the polarization of the radio waves being received. This means measuring the angle of orientation of the radio wave as it propagates through space. It is often difficult to grasp the concept of polarization, but perhaps we can note an analogy with TV and FM signals. Most people have noticed that TV and FM antennas are horizontally placed when mounted outside on a mast. This is because the transmissions being received are horizontally polarized. If you rotated the antenna so that it was oriented vertically you would receive very little signal from the transmitter. Such a rotation is in effect a measurement of the polarization of the received signal. Because you received the maximum signal when the receiver was horizontally placed, you can conclude that the polarization of the radio wave must have been horizontal in your reference system. In the same way the radio astronomer can rotate the (dipole) antenna at the focus of the telescope and find at which angle the received radio emission has a maximum value. He can then calculate the amount of polarization and the orientation of the plane of polarization of the radio wave (see Figure 8). This has a very important bearing on the nature of the source of radiation being observed, since different mechanisms for producing radio waves in astronomical objects produce different polarizations.

Variability

The fourth parameter that needs to be measured is the variation of the received intensity as a function of time. This was not thought to be an important parameter until the late and mid-1960s, when it was shown that most radio sources showed time variations of one sort or another (see Figure 9). The most dramatic of these are the pulsars, whose intensities vary in an extremely regular pulsed manner, switching on and off every second or so.

Therefore, five basic quantities, two positions, one polarization, one giving

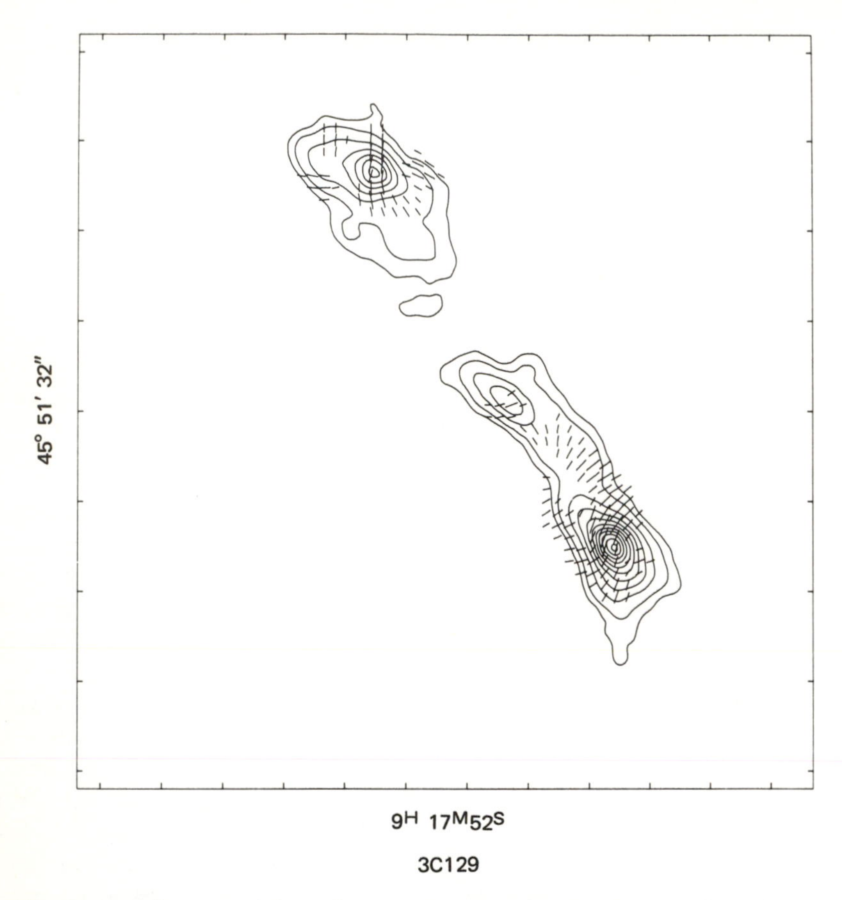

Fig. 8—A radio map of the radio source 3C219 showing the way the polarization measured across the source varies from point to point. The lines indicate the direction of the polarization angle at any given point. (Courtesy NRAO.)

the variation with time, and one defining the spectrum, are needed to fully describe all astronomical objects emitting radio waves, whether they are stars, galaxies, quasars, pulsars, or clouds of gas in the Milky Way. Obviously there are a large number of wavelengths to be examined, and at each wavelength the polarization and apparent position of the radio source might also be different.

Why Radio Telescopes Are Not Shiny

In order to efficiently bounce electromagnetic waves from some surface, the surface has to be very smooth. This is within the experience of most of us, since we are all aware that a smooth pond reflects light very well whereas one covered by ripples does not. In fact the size of the ripples (or irregularities) has to be much smaller than the wavelength we are trying to reflect, otherwise the surface will not make a good mirror. Because light waves have such a short wavelength, optical mirrors have to be made of highly polished glass covered by a silver or aluminum coating, but radio telescopes, which have to reflect only long radio waves whose wavelength is sometimes several inches, need only be made of steel plates fashioned to the right shape to an accuracy of about one-tenth of the wavelength. In other words, the irregularities can be an eighth of an inch or so and this type of surface will not reflect light waves. It therefore does not look shiny to our eyes, but to radio waves it appears perfectly reflecting, or "shiny."

In order to build radio telescopes that can operate at short wavelengths, great care must be exercised to keep the metal plates from deforming when

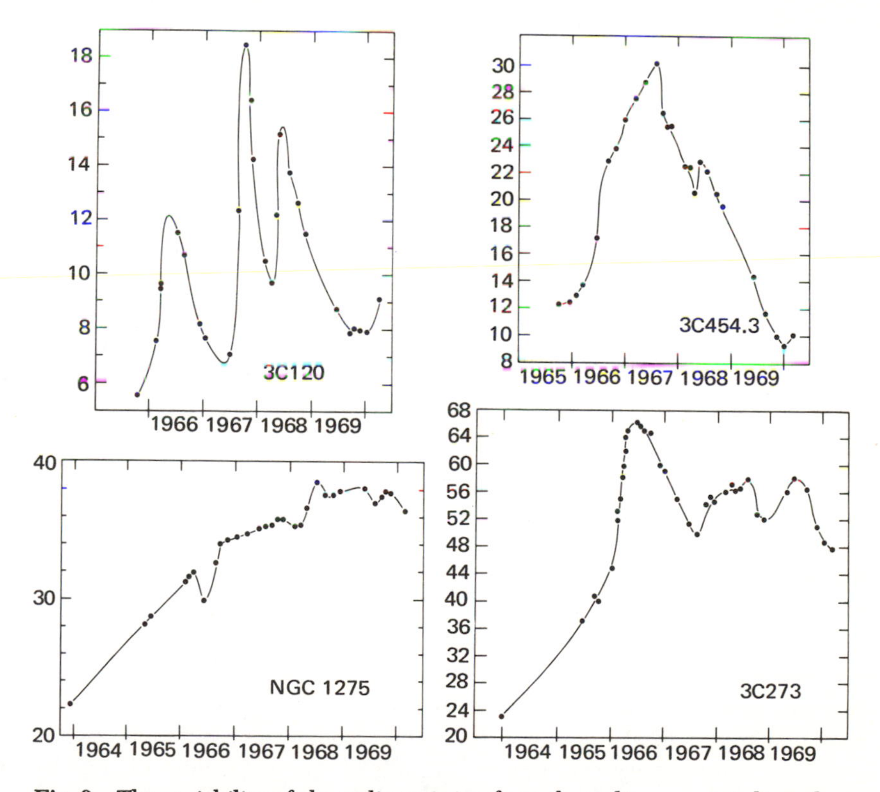

Fig. 9—The variability of the radio emission from the radio sources indicated, as a function of time. Until 1963 it was thought that radio sources were probably steady sources of radio emission, because they were associated with galaxies in many cases. Observations such as these prove that there are very small size hot spots in the objects whose sizes must be of the order of light months or less across. (The intensity scale is in terms of flux units.) (Courtesy NRAO.)

the telescope is tilted, since the gravitational stresses in large dishes can be considerable and will vary as the dish tilts. The largest telescopes on earth are usually not the ones that can operate at the lowest wavelengths, since these have to be engineered more precisely and would be too costly if made too large.

Other Waves in the Electromagnetic Spectrum

Light and radio waves are only two of many wavelength regions of the electromagnetic spectrum. Several other types of radiation, all well known to man, cover shorter and longer wavelengths than those of light. Heat or infrared radiation lies in a wavelength region between that of light and radio waves whereas ultraviolet light has wavelengths shorter than the shortest to which the human eye is sensitive. X rays and gamma rays have successively even shorter wavelengths than those of ultraviolet light.

The Sun sends out all of these radiations, although most of them are prevented from reaching the surface of the Earth because of the presence of a variety of absorbing layers in the atmosphere above the Earth's surface.

Windows into Space

At night we can see the light from stars unless clouds are in the way. The air of the Earth's atmosphere is transparent to light whereas clouds are not. Water drops in the clouds reflect and absorb light as it tries to pass through them. Besides the liquid form of water, the atmosphere also contains invisible water vapor, which hampers the passage of heat, radiation (infrared radiation), and prevents most of the infrared from the Sun, or any other object in space for that matter, from reaching ground-based telescopes. This is fortunate, otherwise we would quickly become unbearably hot in the day, but unfortunate because we cannot easily "see" infrared stars until we get above the atmosphere.

At the other end of the optical spectrum we find ultraviolet light, which is also prevented from penetrating to the Earth's surface, this time because of the presence of molecules of oxygen in the atmosphere. The ultraviolet (UV) radiation knocks electrons out of their orbits around the parent oxygen nuclei, a process known as ionization, which results in the ultraviolet radiation being used up as it passes down through the atmosphere. We say that it is *absorbed*. This is also fortunate because man can not withstand large doses of ultraviolet light either.

A side effect of the ionization is that the electrons so released form an electrically conducting layer well above the main part of the atmosphere, which is called the ionosphere, which, in turn, reflects radio waves of very long wavelengths. The ionosphere therefore makes it possible for radio signals with wavelengths longer than about 10 meters to be reflected around the curvature of the Earth (provided they are generated below the ionosphere, of course) and at the same time prevents long radio waves from outer space from reaching us on the surface of the Earth.

Between the wavelengths of about 10, or sometimes as much as 30 meters, and the lower limit, of approximately half a centimeter, the Earth's atmosphere and ionosphere are perfectly transparent to radio signals. We speak of this as being a "window" in the electromagnetic spectrum through which we can observe the distant universe. Another window through which we can look out at space is of course the optical window, blocked at one end by water vapor in the air and at the other end by oxygen.

The optical astronomer peers through the optical window whereas the radio astronomer observes through the radio window. X-ray, gamma-ray, ultraviolet, and infrared astronomy can be successfully carried out only above the atmosphere with rockets, satellites, and balloons carrying the telescopes for these wavelength regions.

The remainder of this book is all about the things we "see" through the radio window into space.

Direction Finding in Radio Astronomy

A radio astronomer uses a single radio telescope to pick up radio signals coming from one direction at a time. Unlike the human eye the dish can look at only one spot in the sky at a time. We are completely used to seeing an image that covers quite a large area in front of us, but this is a tremendous luxury compared to what the radio astronomer can see with his big dish. Imagine what sort of view we would have of the world around us if, while looking at (for example) a small boat sailing on a lake surrounded by beautiful mountains, we could only see the boat, or perhaps only a part of the boat, at any instant. In order to see the whole scene we would have to scan slowly the whole area to try to store the information until we had completely surveyed the region, and only then could we put all the pieces together. This is what the radio astronomer usually does. The eye has many sensing elements, which allow this process to be completed very nearly instantaneously, but the radio telescope, with its single antenna, has only a single sensing element, and therefore scanning every part of the sky may take days or years. The process can be speeded up by the installation of more than one antenna at the focus of the dish or by the use of several dishes in the right way, but this never approaches the sophistication that we are used to in the human eye.

Not only can the radio telescope be pointed at only one spot at a time, but what it sees is never sharply in focus. Instead everything is rather blurred. This has nothing to do with focusing the telescope as we do with optical telescopes or binoculars; it is the result of dealing with radio waves rather than light waves. It is a question of the ability of the telescope to "resolve" details in angle. The *resolving power,* as it is called, depends on the size of the telescope and on the wavelength used, in the following way. The larger the diameter of the telescope, the smaller the structure it can resolve; that is, see clearly, and at the same time the shorter the wavelength of the radiation to which it is tuned, the better it is able to resolve small-scale structure.

The human eye, which is tuned to a wavelength of approximately 50-millionths of a centimeter, can see details of about 20 seconds of arc in size, assuming that the pupil has a diameter of about half a centimeter. The 200-inch optical telescope at Mount Palomar would have a capability of resolving details down to one-fifteenth of a second of arc in size, if atmospheric conditions permitted it. For comparison purposes, a penny placed at a distance of one kilometer has an angular size of only half a second of arc. The formula used to get these numbers is: the resolution of the telescope (or eye, etc.), in radians, is given by the wavelength used divided by the diameter of the collecting device (eye, mirror, dish).

Let us see what sort of numbers apply to radio telescopes. First of all, the diameter of the largest movable dish is about 100 meters (300 feet), and the wavelengths typically used are of the order of 20 centimeters. This produces a resolution capability of about only 10 minutes of arc (called the beamwidth; see Figure 10), which means that details smaller than this are simply lost and everything in the sky at which the telescope is pointed appears to be at least this big. We might say that everything appears blurred. The human eye is

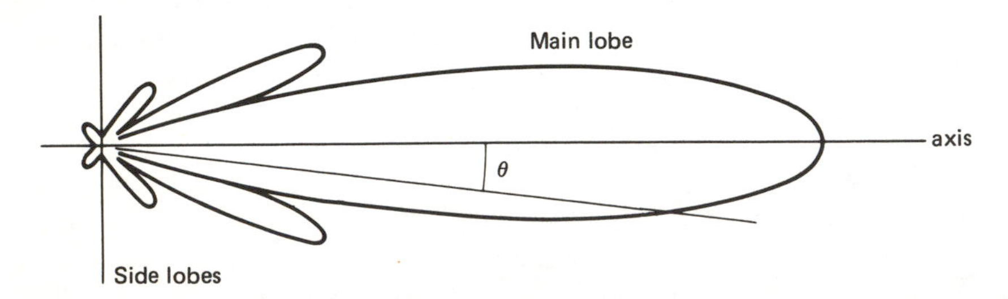

Fig. 10—The radiation pattern, or reception pattern, of an antenna, showing a cross section of the pattern in a plane containing the axis of the antenna. Such a diagram is often called the polar diagram of the antenna. The beamwidth of the antenna would be 2θ if the intensity of the received radiation has half its peak value at a distance θ from the axis. A map of a pointlike radio source made with this antenna would show the radio source to have an apparent size of 2θ rather than to be pointlike, which reflects the limitations imposed by using a single antenna with this polar diagram.

therefore a much better instrument for seeing (resolving) small-scale structures in space than a radio telescope is. In order to obtain a resolution (beamwidth) at 20 centimeters' wavelength equal to that of the human eye, the radio telescope would have to be about 200 kilometers in diameter!

Radio astronomers can actually achieve such a capability by using two radio telescopes separated by that distance or more, in order to obtain very fine scale information on radio sources in space.

An aspect of this resolution problem can easily be demonstrated by using a portable radio. If one picks up a fairly distant radio station, he must turn the portable until the strongest signal is picked up. The portable antenna has its own *response pattern* to incoming radio waves, just as a radio telescope does. Now consider how accurately one could pinpoint the direction of the transmitter being picked up. First of all the radio can be rotated through several degrees at least without the possibility of detecting the change in the strength of the broadcast. Now let us say that the transmitter is located 60 miles away. At that distance an uncertainty of one degree in the direction determined for the transmitter is equivalent to about one mile, which is not much use if one wants to describe exactly where the transmitter is and at what height above the ground it might be. A portable radio would, therefore, be no good for finding exactly where the transmitter is located. One would need a much larger antenna in order to do that with any accuracy.

This then is the problem that faces the radio astronomer using a single big dish. He might know that radio waves are coming from some direction in space, but the position of the radio source could equally well lie anywhere within a box drawn on the sky whose sides might be 20 minutes of arc in size. One need only look at any photograph of the sky taken with a large optical telescope to see what an enormous number of stars, and sometimes galaxies, are included in such an area. Clearly it is impossible to state which object is responsible for the radio signals received unless there were to be some very obviously peculiar object at that position which was a likely candidate. We shall see later that in some cases very obvious objects can be recognized, even when only the position estimates from single radio telescopes are available. These are often old supernovae or some nearby galaxy or incandescent gas cloud in the Milky Way, but more often than not the identification of the majority of radio sources, such as quasars, depends on using very different observational methods and very different telescopes.

Protected Bands in the Radio Wave Region

Before we start describing the discoveries of radio astronomy, it is of interest to point out that often the radio astronomer must struggle to make his observations, because there are transmitters on Earth that cover nearly all the radio wave part of the spectrum. TV, and FM and AM radio stations are obvious examples, but there are also military communications channels, radar transmitters, airline control channels, radio beacons, taxi cabs, and "hams" on the air. All these signals produce unwanted "interference" for the radio astronomer trying to pick up weak signals from outer space. Other sources of interference are motor cars, airplanes, tractors, and a host of electrical devices.

For these reasons radio observatories are hardly ever located near towns or cities (with some notable exceptions). To enable the science of radio astronomy to progress at interference-free wavelengths, several wavelength regions have been allocated by international agreement for the exclusive use of radio astronomers, and no one may transmit at those wavelengths for any reason. This works fairly well in practice, especially when the radio astronomer is able to predict which bands are of critical importance to him, such as the 21-centimeter region at which cosmic hydrogen clouds transmit, but the recent discovery of many molecules in space, each transmitting at some different wavelength means that the radio astronomer has to search for some signals in wavelength regions reserved for other uses. This makes his task so much more difficult.

CHAPTER 2

THE BIRTH OF RADIO ASTRONOMY

In 1933, a Bell Telephone Laboratory engineer named Karl Jansky tackled the very practical problem of locating the sources of interference hampering transatlantic telephone links. This piece of research turned up the startling fact that the Milky Way itself was an important source of interference. This discovery was followed by the work of a radio amateur, or "ham," called Grote Reber who built his own radio telescope, not unlike the type still commonly used today, and mapped the emissions being produced in our galaxy, which we call the Milky Way. Thus radio astronomy was born, and it is now one of the most exciting research fields in which to work.

Tracking Interference

The beginning of radio astronomy on Earth can clearly be connected to Karl Jansky. He constructed an antenna that could be rotated on a set of wheels to point to any part of the horizon (see Figure 11). In this way he was able to recognize that distinct signals were being produced by thunderstorms in the immediate vicinity of his antenna as well as by those much farther away, well beyond the horizon. In addition to the lightning flashes, which produced very distinctive cracking sounds in his receiver, Jansky also noticed that a distinctive hissing sound was almost always present, although it varied in intensity during the day with a period of not quite 24 hours. In fact it seemed to vary uniformly, going from a peak through a minimum and back to a peak with a period of 23 hours and 56 minutes.

In order to explain this time pattern Jansky turned to astronomy for the solution. The characteristic time of 23 hours and 56 minutes is the so-called sidereal day, which is the time the Earth takes to rotate once on its own axis when viewed relative to the stars. The 24-hour day we normally go by is the

Fig. 11—Karl Jansky with his antenna that first picked up radio signals from the Milky Way. (Courtesy NRAO.)

time taken for the Earth to rotate once on its axis when measured relative to the Sun; that is, the time taken for the Sun to pass from being directly overhead on one day to the time it is directly overhead the next day. The source of a hissing sound Jansky heard with his equipment was unlikely to show the sidereal periodicity if it was associated with the Earth or the Sun in any way. It was far more likely coming from a region out among the stars.

It did not take Jansky long to realize that the hissing was the Milky Way moving across the sky, producing the varying radio signals. For some reason the Milky Way itself was transmitting radio waves and he had accidentally picked them up with his equipment.

Let us briefly explain what the Milky Way is and why it looks the way it does. The Milky Way is the name given the bright band of stars that can be seen on most clear nights, and it is particularly dramatic in the southern skies. The increase in brightness is due to the presence of many thousands of stars in this band, whereas in other directions in the sky the number of visible stars is considerably less. The reason for this particular scattering of stars is that we are located inside a disk- or wheel-, shaped galaxy, which is an enormous conglomeration of stars and gas that drifts as a coherent and stable entity through space. Because we are located inside this disk, we see more stars when we look out along the disk than when we look out at right angles to it.

The term Milky Way is often used as the name for our galaxy, which is but one of many thousands of millions of galaxies in the universe. Remember that each galaxy contains thousands of millions of stars, and when we look at the Milky Way at night we can see only the nearest few thousand stars in our own galaxy.

Jansky discovered that the Milky Way was transmitting radio waves, although he had no idea at the time why that should be so. Obviously the Bell Laboratory engineers could not do anything to prevent this cosmic source of radio signals from interfering with the telephone lines, so Jansky was taken off the project and given something else to work on.

Radio astronomy would not have continued its growth for quite a while had it not been for the enthusiasm shown by a radio amateur, or ham, by the name of Grote Reber of Wheaton, Illinois. Reber happened to come across the technical reports describing Jansky's discoveries and decided that he, Reber, should follow them up to try and find out more about the nature of the radio signals from the Milky Way. For this he needed a larger antenna with better directional capability than the one used by Jansky. Reber realized that he could most easily achieve this by constructing a dish-shaped reflector and placing his antenna at the focus of the dish. His radio telescope was in fact exactly of the form that all subsequent big dishes have had and yet he constructed his 32-foot diameter dish single-handed in his own back garden. Nowadays many keen amateur astronomers are still doing this sort of thing, but try to look back and picture the surprise of people not aware of what Reber was doing when they saw the strange structure appearing in his garden. Nothing like it had ever been constructed on Earth. Also try to envisage the reactions Reber must have gotten when he explained, back in about 1935, that the device would allow him to pick up radio waves from space. The story goes that since the dish had a hole in the bottom for allowing rain water to drain out, people suspected that it might, in fact, be a rain-making device! Even nowadays it is not difficult to imagine that some radio telescopes are simply enormous rain gauges.

Nevertheless Reber persevered and succeeded in confirming what Jansky had found—that the Milky Way was an enormous radio transmitter and that the sound produced by it, the steady hiss, was in fact similar to the hiss produced by the radio sets themselves. What was being received was radio noise from space. In addition, the improved resolving power of Reber's antenna allowed him to discover that stronger radio signals were coming from the apparent center of the Milky Way system than from other points along the bright band and that several other regions were also emanating stronger radio signals. These points were subsequently referred to as radio stars and the nature of most of them was not understood for many years after Reber's discovery. None of them turned out to be true stars and are now called by the more general name *radio sources.*

Reber subsequently produced many maps of the way the strength of the received radio waves from the Milky Way varied over the sky, and his earliest results were finally published, not without considerable difficulty and subsequent skepticism from many astronomers in the scientific journals in the early 1940s. Reber's telescope has now been reerected at the National Radio Astronomy Observatory in Green Bank, West Virginia, and is on public display and is still sometimes used.

Radar Jamming by the Sun

As is so often true of pioneering work in science, Reber's work went largely unrecognized for some time, which was partly due to the onset of World War II, which prevented those who appreciated what an important discovery had been made from doing anything more about it. On the other hand, scientific development of radar equipment during the war lead to another discovery in radio astronomy, which was that the Sun was also a source of strong radio signals. This discovery resulted from the work of the radar researchers in Britain who found that severe jamming signals, first suspected as having been produced by the enemy, were in fact coming from the Sun. The interference was always worst around dawn, which was reasonable since most of the radar sets in Britain were pointed toward Europe, the direction from which the Sun rose. Furthermore, the scientists discovered that the signals were strongest when sunspots were visible on the Sun's surface, which meant that the radio emissions had to be associated with the spots themselves.

This discovery, and the work of Jansky and Reber, was seriously followed up only after the war had ended, and when it did many scientists found that suddenly plenty of equipment became available for radio astronomy. War surplus equipment included a supply of radio receivers, which were necessary to tune in to the cosmic radio broadcasts, and more importantly there were large quantities of the dish-shaped antennas. It turned out that some of the most useful radio telescopes for nearly 10 years after the war ended were the German "Wurzburg" radar antennas (Figure 12), which are still used sometimes at both the Dutch (Dwingeloo) and the English (Mullard Observatory at Cambridge), radio astronomy observatories.

The centers of radio astronomy research in Europe that subsequently grew up were therefore based around those men who had worked in radar research, who had managed to learn about Reber's work, and who had also been able to salvage working radar telescopes and receivers. Some of the more famous observatories started in this way were Dwingeloo in Holland, using a Wurzburg antenna; the Cambridge group under Martin Ryle, also using a Wurzburg; the Jodrell Bank station of the University of Manchester

Fig. 12—The old Wurzburg antenna, a German radar dish from World War II, used for making some of the first radio astronomical observations in England as well as in Holland. This one is in Cambridge, England. (Courtesy Mullard Radio Astronomy Observatory.)

under Bernard Lovell, using British radar equipment; and a group at the Royal Radar Research Establishment at Malvern under J. S. Hey.

At the same time a group in Australia got started and did considerable work in solar radio astronomy during subsequent years. Not much radio astronomy was started in the United States with the exception of the work at the Naval Research Laboratory in Washington, D.C. Let us briefly consider some of the earliest results from these laboratories.

Radio Signals from Gas Clouds

Despite the war, an important prediction was made at a scientific meeting in occupied Holland in 1944, by a young student named Henk van de Hulst, who calculated that hydrogen atoms should act as tiny radio transmitters (this will be explained in Chapter 7) and that the radio signals produced should have a wavelength of 21.2 centimeters. He suggested that this signal should in fact be observable if suitable radio receiving equipment were constructed and the radio sets were tuned to this wavelength. After the war ended three research groups set off in a great hurry to try to discover these signals, and all three succeeded, more or less simultaneously, in 1951, although the Dutch group would have been the winners if they had not suffered a disastrous fire that destroyed much of their equipment at a crucial stage in the race. The American group consisting of Ewen and Purcell were in fact the first to detect the radio signals from the hydrogen gas existing between the stars. This gas was made observable to man for the first time via radio astronomy, as it is not detectable in other ways. Hydrogen is the most basic element in the universe and is used to build stars in which all the other elements are subsequently formed. The study of the distribution of the hydrogen in the Milky Way is therefore a very important part of radio astronomy research today.

Similar competition between various observatories still exists and recently the radio astronomy world has been witness to at least two major "rat races"—the race to discover as many new pulsars as possible and, more recently, the race to discover as many new molecules as possible. Both these races are discussed later.

Early Days at Jodrell Bank

About 20 miles south of Manchester, England, in rural Cheshire, is located one of the most famous radio telescopes in the world. Much of its fame rests on the enormous public exposure it received during the early days of Sputnik and the space race, and its director (now Sir) Bernard Lovell was well known as an outspoken commentator of the space program. He directs the Nuffield Radio Astronomy Laboratories at Jodrell Bank, which are a part of the physics department of the University of Manchester.

Lovell was one of those scientists who had acquired radar equipment after the conclusion of the war and in 1945 had tried to start measurements in Manchester. This turned out to be a fruitless endeavor because of the tremendous amounts of interference generated by electrical equipment in the city, in particular by tram cars. Fortunately the university had a botanical research station at Jodrell Bank and there Lovell was allowed to park his trailers of equipment.

In his first experiment he wanted to try to pick up radar echoes from the trails that cosmic rays were thought to make in passing through the Earth's atmosphere. Cosmic rays are fast-moving particles known to be traveling through much of the space that fills the air around us, passing right through our bodies without our ever noticing them. These particles were thought to produce ionized trails in the atmosphere, which means that they stripped the electrons from the atoms in the atmosphere, after colliding rather violently with them. Since such ionized trails are electrical in nature, they might be expected to reflect radar signals. During the war, Lovell and other radar workers had noticed the presence of sporadic signals on their screens, which were thought to be due to cosmic ray trails. Since the cosmic ray particles travel at nearly the speed of light, the phenomenon was expected to be very short-lived, so this explanation for the spurious echoes did not seem unreasonable.

Using his equipment at Jodrell Bank, Lovell was to make a surprising discovery soon after he started working. One night he picked up thousands of echoes of the type he had been planning to study. The usual rate was thought to be only one or so an hour. On that particular night an accomplished amateur astronomer was collaborating with him in his research, watching the sky outside the trailer for meteor trails. Lovell had for some time had the suspicion that the echoes might be produced by the echoes from meteor trails instead of cosmic rays, and he solicited the aid of the amateur astronomer who was well aquainted with counting meteors. It turned out that while the thousands of echoes per hour were being recorded, a meteor shower was occurring overhead and Lovell reports that the skies were literally ablaze with a dramatic meteor shower known as the Giacobinids.

Meteor radio astronomy was thus born, and for the next 10 to 15 years research in this field was to be the main scientific thrust at Jodrell Bank. As the meteor burns up in the Earth's atmosphere, it causes ionization of the gas along its path, and it is from these trails that the radar echoes were picked up and not from cosmic ray trails. The latter have still not been detected and are unlikely to be, although in the mid-1960s, scientists at Jodrell Bank did pick up direct radio signals produced by the cosmic ray showers as they struck the Earth's atmosphere. Examination of the radar echoes from meteor trails gives important information on the paths that the meteors were following at the time and these can be traced backward to give an idea of the direction from which the meteors originally came.

While this research was continuing, Lovell and his assistants constructed a 218-foot-diameter dish with a surface made of chicken wire, and this dish was used for some early radio astronomical observations at Jodrell Bank. It was Lovell who first managed to construct a really large steerable radio telescope. The 250-foot dish (Figure 13) was completed in 1957 just before Sputnik 1 was launched. This telescope was completed only as a result of some incredible perseverance on Lovell's part, and the story of how he made this actually come about is described in his fascinating book *The Story of Jodrell Bank*.

The launching of Sputnik 1 served to bring radio astronomy as well as the space program and the space race itself into the public eye because it turned out that the 250-foot radio telescope at Jodrell Bank, when equipped with a simple radar transmitter designed for meteor work, was the only radar on

Fig. 13—The Mark IA radio telescope at Jodrell Bank in England. This is the restructured version of the older Mark I, with a better surface and focus laboratory installed and a more solid support behind the dish. The diameter is 250 feet. The lowest closed-in space in the support tower is some 120 feet above ground. (Courtesy Nuffield Radio Astronomy Laboratory.)

Earth capable of picking up echoes from the Sputnik carrier rocket. This apparently upset much strategic thinking in the United States! Radio signals from Sputnik itself were easily picked up by any short-wave receiver and it certainly required no large radio telescope to make that possible. Nevertheless Jodrell Bank and Lovell became firmly associated with the space age in the eyes of the British press and throughout the world. Sometimes Jodrell Bank was described, in foreign newspapers, as the British satellite tracking station, much to the delight of scientists at rival establishments!

So it was to be for many succeeding years; every time a major space venture was undertaken the radio telescope at Jodrell Bank received more and more publicity. Some dramatic events did take place at Jodrell, many associated with the Russian space probes and the rockets that went to the Moon. Jodrell Bank was the only observatory in a position to monitor the efforts of the Russians in landing their first spaceships on the Moon. It was possible to confirm or deny Russian statements concerning the maneuvers of their rockets, since they freely gave Jodrell Bank much information concerning the wavelengths at which their satellite or space probe equipment would be transmitting. If any of the radio telescopes belonging to the U.S. Armed Forces were performing similar monitoring tasks, no one was saying, so it was to Jodrell Bank that everyone turned to learn about the latest Russian venture into space.

A most dramatic climax to the use of a radio telescope designed for astronomical research in the space program came with the reception, at Jodrell Bank, of the first photographs transmitted from the Moon's surface. While all this excitement was taking place in Britain, the United States had entered the radio astronomy field in a very systematic way by setting up the National Radio Astronomy Observatory (NRAO) under the trusteeship of Associated Universities, Inc., which is also responsible for running the Brookhaven National Physical Laboratories in Brookhaven, New York. The NRAO is currently the most productive and successful radio observatory in the world, although that is not meant to detract from the good work being done elsewhere. Before we start on the tour of the universe as revealed by radio astronomy, let us briefly describe how the NRAO, which operates for the benefit of all U.S. radio astronomers and many from Canada, Europe, Australia, and other countries, is run.

The U.S. National Radio Astronomy Observatory

The National Radio Astronomy Observatory (NRAO) in Green Bank, West Virginia, exists as a national facility, because the astronomical community realized, in the late 1950s, that the cost to a single university department to own and operate a major piece of scientific equipment like a large radio telescope would be prohibitive. The National Science Foundation therefore gave a group of universities the go-ahead to oversee the construction and development of a national center for radio astronomy in the same way that such a center had been set up for physics in Brookhaven, New York. This group of universities, operating as Associated Universities, Inc. (AUI), operates Brookhaven National Physical Laboratory and the NRAO with funds provided by the National Science Foundation (NSF).

At present the NRAO in Green Bank has one of the finest high-accuracy dishes in the world in the form of the 140-foot fully steerable telescope and one of the largest movable dishes, the 300-foot tiltable telescope. In addition, three 85-foot-diameter dishes operate as an interferometer to produce very accurate pictures of radio sources.

At Kitt Peak, near Tucson, Arizona, the NRAO owns and operates a very accurate 36-foot radio telescope (Figure 14) for making measurements at very short wavelengths of the order of a few millimeters to a few centimeters.

Fig. 14—The horn antenna (the square-shaped object) mounted at the focus of the 36-foot-diameter radio telescope at Kitt Peak, Arizona. (Courtesy NRAO.)

When it was completed in 1967, very few objects in the universe could be observed with this telescope, not because it was a bad telescope, but simply because there were few radio sources radiating appreciably at these wavelengths. This led to some pessimism about the future usefulness of that telescope, but since then, the startling discovery of molecules in space and the fact that many of them radiate in this short wavelength region has meant that most of the time on the 36-foot is now spent in searching, with considerable success, for more molecules. Without the 36-foot dish much molecular astronomy would be impossible to perform.

The main scientific offices and the computing center for the NRAO are located in Charlottesville, Virginia, which, with a population of 38,000 and about 35 to 45 astronomers in the city must have one of the highest densities of astronomers per capita in the world!

About 70 percent of all the radio astronomical papers published in the United States in 1970 were based on observations made using the NRAO facilities. The telescopes may be used by all qualified astronomers and their research students, provided they have reasonable research programs planned. Not only U.S. astronomers, but astronomers from the U.S.S.R., India, Australia, and most of the European countries, have worked at Green Bank.

The telescopes are operated by professional telescope operators, and the receiving equipment is mainly constructed and maintained by a highly qualified electronic engineering staff, who develops much of the special equipment required, most of which is not yet commercially available (see Figure 15).

The Observatory in Green Bank is open to tourists during the summer, and tours of the site as well as informative films and display boards give the visitors some idea of the work being done there.

Fig. 15—A view of the control room of the 300-foot-diameter radio telescope of the NRAO in Green Bank, West Virgnina. Here two radio astronomers are discussing a problem with an electronics engineer. (Courtesy NRAO.)

CHAPTER 3

THE RADIO SUN AND PLANETS

Radio Signals from the Sun

To radio "eyes," the Sun does not appear to shine with a steady glow. Enormous storms, lasting several days, occur near sunspots, causing intense bursts of radio emission to radiate into space. Giant explosions cause the solar atmosphere to vibrate and clouds of particles stream out toward the planets, and at the start of their journey they emit intense short-lived bursts of radio waves that completely drown out the normal radiation from the quiet Sun.

The sunlight we see with our eyes comes from the surface layer of the Sun, called the photosphere, whose temperature is about 6000° Kelvin. Above the photosphere is another layer, which is the atmosphere of the Sun. This is the *corona,* whose temperature is of the order of millions of degrees and which can be seen visually only during a total eclipse of the Sun. However, radio astronomers can always "see" the corona with their radio telescopes, since it is the hot solar corona that transmits strong radio signals, and not the cooler photosphere.

The 6000° Kelvin temperatures in the solar photosphere cause the particles there, in particular electrons, to rush around at very high speeds, and collisions between them produces the light we see. The wavelength at which signals are sent out by such a process depends primarily on how many particles there are in every cubic centimeter of space (the density) as well as the temperature. The higher the density, the shorter the wavelength of the radiated signals; the lower the density, the longer the wavelength radiated.

This is also true for the solar corona, but the higher we are above the visible surface of the Sun, the fewer particles there are, and therefore the signals that are generated there have a much longer wavelength. They are in fact radio waves that can reach Earth. We cannot, at this point, go into the question of why the solar corona is so much hotter than the lower visible surface (the photosphere), but it does mean that radio astronomers observe a

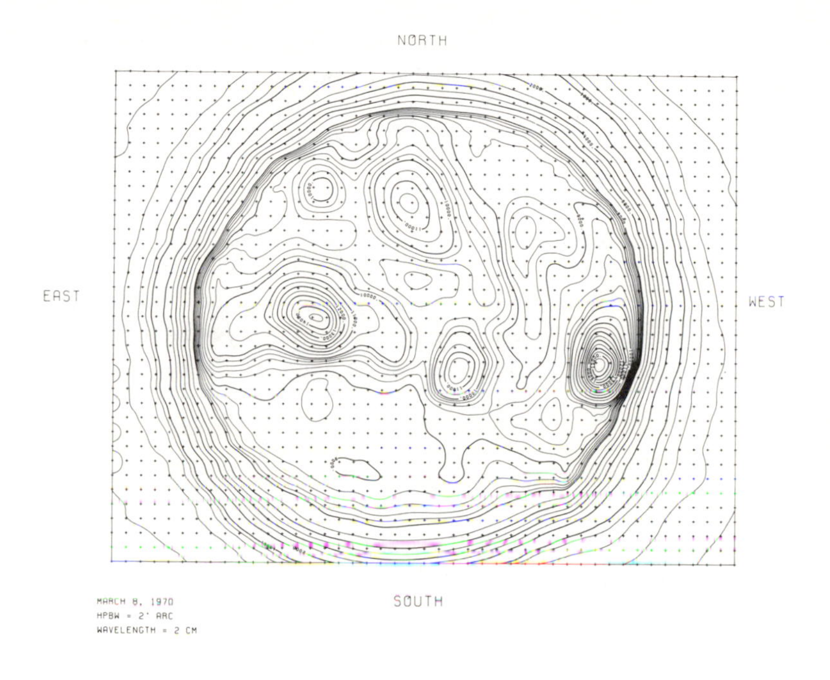

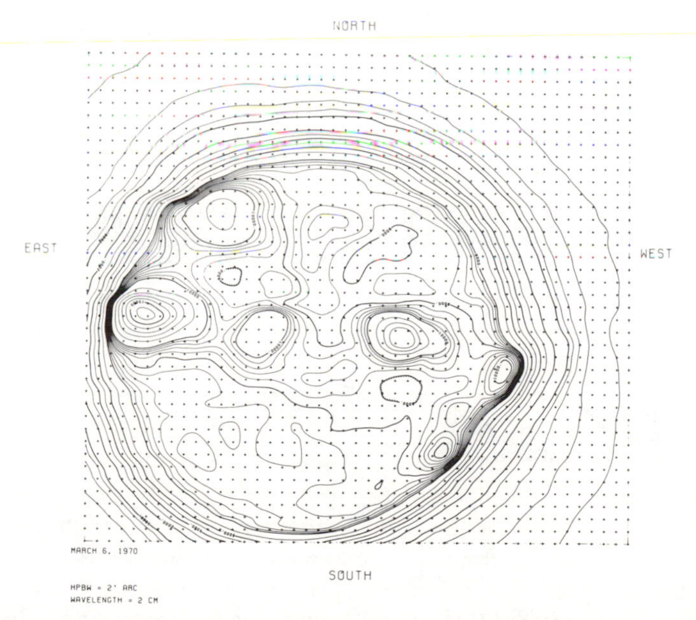

Fig. 16—Two radio maps of the Sun made at a wavelength of 2 centimeters, showing the movement of centers of radio activity, usually associated with sunspots changing their structure as well as moving across the disk as a result of solar rotation. (Courtesy NRAO.)

much hotter sun than we see. The Sun's heat (infrared radiation) also originates from the surface layers and not from the corona. It is true that if the density of the corona were high enough and the temperature stayed at a million degrees Kelvin, then Earth would quickly be roasted by the heat radiated.

Radio signals produced by the corona produce a steady hissing sound in a loudspeaker, and no matter to what wavelength the radio set is tuned this hiss can be heard. Its strength may vary only slightly from day to day, and if there is no visible activity on the Sun, such as sunspots or flares, this steady signal might vary, at most, only slowly over a period of 27 days (the time for the Sun to rotate once on its own axis). A map of the quiet Sun is shown in Figure 16.

Solar Storms and Explosions

Occasionally, and mostly at sunspot maximum, dramatic occurrences are visible on the surface of the Sun. These are the so-called flares that are visible as very bright increases in light in small regions on the Sun's surface; they are

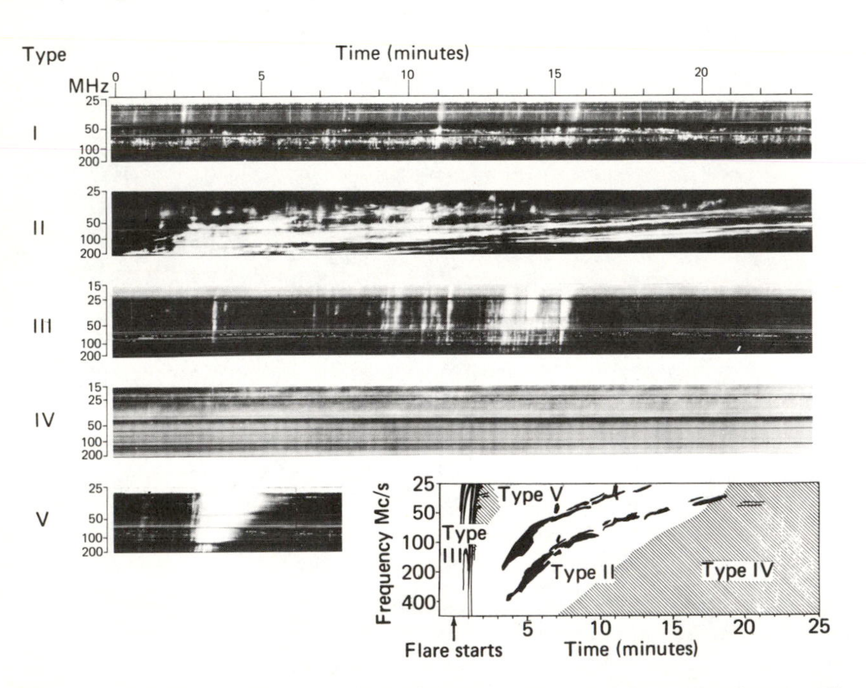

Fig. 17—Actual examples of all five basic types of solar radio bursts. The brightening in the photographic records indicates an increase in the amount of radio emission being received at any particular time and frequency. The change of frequency with time for some events indicates that the source of the radio waves is moving up through the solar corona. A schematic diagram summing up these data is also shown. (Courtesy C.S.I.R.O., Division of Radiophysics.)

produced by sudden violent explosions. These explosions send clouds of particles shooting upward into the corona, and the path of these clouds can be followed very closely by radio astronomers for the following reason. When such a cloud of particles passes through any given layer in the corona, it disturbs the region in its immediate vicinity in such a way that the electrons that are normally there will start to oscillate in unison at a definite rate determined by the density of the material. These oscillations produce radio waves, which we can pick up on the Earth with suitable radio sets and radio telescopes pointed at the Sun. As the cloud of particles moves upward through the corona it generates radio waves of successively longer and longer wavelengths, and on Earth this change of wavelength can be monitored and interpreted as the motion of the particles in the solar atmosphere (see Figure 17).

The reason for the change of wavelength is that as the disturbance moves upward it encounters regions of successively lower density—and the period of the oscillations is determined by the density, becoming slower for lower densities. This means that the longer wavelength emission comes from higher in the corona. Such oscillations of a hot gas, as are found in the corona, are called *plasma oscillations*.

This type of radio signal, resulting from a flare, is called a *solar burst*. There are several types of bursts associated with flares and yet other types of bursts associated with sunspots. The one we have described is a burst of type III; the numbering reflects the order in which they were discovered and little else, as is so often the case with astronomical notations.

While the particles described above generate plasma oscillations, other particles, usually electrons, are accelerated to very high speeds by the explosions, which are the flares themselves, and these accelerated particles in turn radiate radio signals because their motion is controlled by the magnetic fields associated with the flares. The high-energy particles are forced to spiral around the lines of magnetic field, and in this process they lose energy in the form of radio waves, which are detectable on Earth as type V bursts, which cover a wide wavelength range at any given time and which might last many tens of minutes. These are unlike the type III bursts, which are only short-lived and cover a very narrow wavelength range at any instant. The type V burst covers a large range of wavelengths because the wavelength generated does not depend on the density of the electrons involved, but only on the strength of the magnetic field and the speed of the electrons, both of which have a large range of values in the immediate neighborhood of the flare.

Very often, after the initial explosion has subsided, a secondary cloud of particles is ejected from the location of the flare, an ejection not usually visible optically, and this process also generates two types of radio bursts similar to those described above. This second cloud of particles moves at only about one-tenth the speed of the first, and therefore we see these type II bursts moving slowly in wavelength as the cloud moves up and out of the corona into space. The associated type IV burst again covers a wide wavelength range and might last for half an hour before all the energy is radiated away.

Some of the clouds of particles emitted by the Sun in this way ultimately reach Earth, provided they were headed in the right direction to start with, and then they produce such dramatic happenings as the Corona Borealis and radio communication blackouts.

Storm Bursts

The better-known solar events called sunspots also produce characteristic radio bursts, called storm bursts or bursts of type I. These are very short-lived intense pips of radio emission, each lasting only a fraction of a second, although thousands of these might be emitted every hour and the noise storm (a string of bursts) may last for days.

Sunspots are associated with locally very strong magnetic fields and there are many high-speed electrons produced which, as in the case of the type IV and V bursts above, spiral around the magnetic field lines and in this case, due to the different field strengths and energies of the particles, the emission is observable as a series of bursts, possibly with an underlying enhancement of emission. In the case of weak sunspot activity, only this enhancement of the total radio emission produced by the Sun is observed, without any associated burst activity.

"Photographing" Solar Radio Bursts

Paul Wild, an Australian who has for many years been an active solar radio astronomer, has constructed a large circle (diameter 2 miles) of 96 radio telescopes, each 13 meters in diameter, especially designed to observe the solar radio bursts in a way that allows him to obtain a movie of the way in which the clouds of particles described above move out from the Sun (see Figure 18).

All of the telescopes are connected to a central computer and this produces a single radio "picture" of the Sun by combining in the right way the signals from all the telescopes. Such a combination of radio telescopes is called an interferometer and many varieties of interferometers exist throughout the world today. The radio picture of the Sun at any instant is displayed on a television screen, and by photographing every new picture generated a

Fig. 18—Two of the 13-meter-diameter antennas that make up the Wild circle at Culgoora in Australia. The circle is used as an interferometer for studying solar radio waves. (Courtesy of C.S.I.R.O., Division of Radiophysics.)

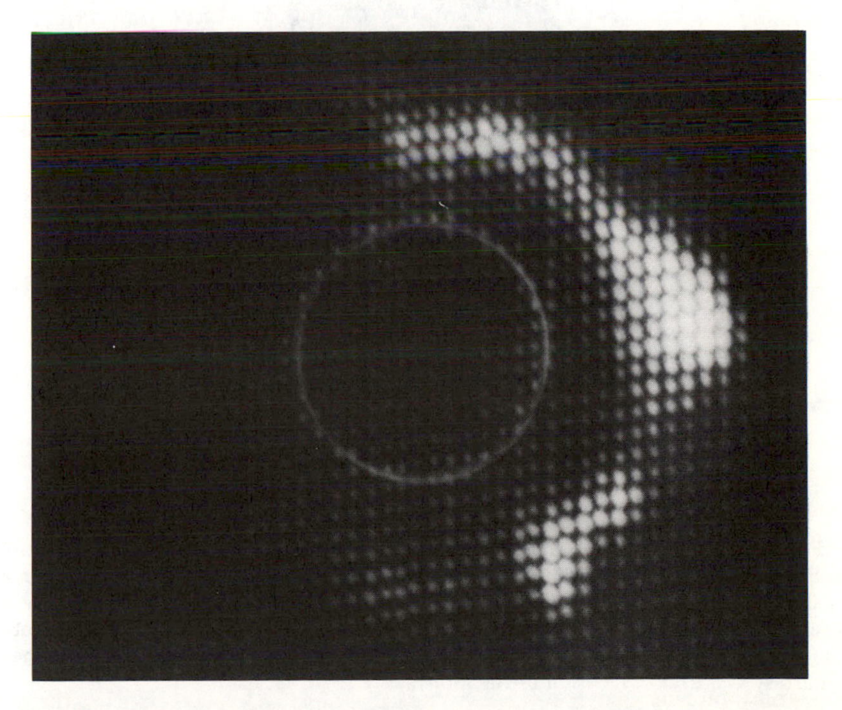

Fig. 19—A "photograph" of a type II burst which was produced by a proton flare behind the west limb of the Sun. (Courtesy C.S.I.R.O., Division of Radiophysics.)

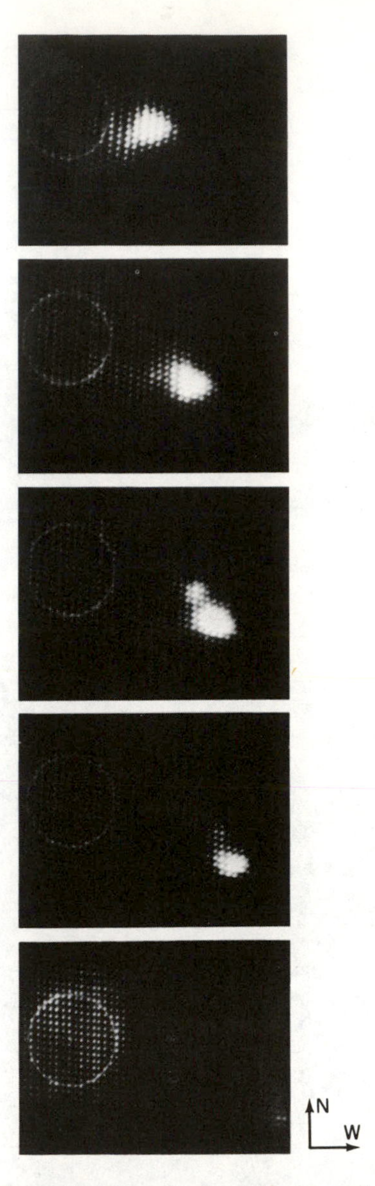

Fig. 20—An example of the type of data produced by the Culgoora radioheliograph. In this photograph of the radio signals being received from the Sun, after display on an oscilloscope of course, a moving type IV burst can be seen. It is produced by an ejected blob of hot gas as it moves outward from the Sun, whose optical extent is indicated by the circle. This burst was called "Westward-ho" by radio astronomers. (Courtesy C.S.I.R.O., Division of Radiophysics.)

movie is obtained, which then gives dramatic evidence for the motion of clouds of particles out from the Sun. This motion is visible as a bright patch on the picture moving away from the Sun, and other events such as short-lived bursts appear as sudden flashes on various parts of the Sun's surface. A circle, representing the visible solar disk, is drawn on the screen, so that one can quickly tell where on this disk the radio event has occurred (see Figures 19 and 20).

This circle of radio telescopes, known as the Wild circle, simulates a radio telescope of a diameter of a mile. Clearly it is impossible to construct a dish with a diameter of a mile, but this circle of telescopes serves nearly as well. The resolving power is determined by the diameter, but clearly not all the radio emission reaching the area of the circle can be collected by the system, since there is much empty space left over. However, the computer makes the best use of the data collected and manages to produce a good radio picture of the Sun, partly because the Sun is such a strong radio source and partly because the radio astronomers know more or less what is expected. In other words they are not searching for new weak radio sources with this array of telescopes, because then the lack of a completely filled area produces problems that we shall touch upon later.

The Planets

Black-body Emission

Even the planets in the solar system send out radio waves, and the most dramatic aspect of these, the radio bursts produced by Jupiter, was discovered accidentally.

There are many ways in which radio waves can be generated in astronomical bodies but basically every object in the universe, no matter what its temperature, as long as it is not at absolute zero, will radiate signals at all wavelengths in a way determined only by its temperature. This is the result of collisions and near collisions between electrons in the material, whether it be a solid object like the Earth, at 300° above absolute zero, an interstellar gas cloud at 20° above absolute zero, or the solar corona at a million degrees above absolute zero. It is possible to measure the intensity of this so-called thermal emission at any wavelength one chooses. If one chooses to measure in the infrared, then one effectively uses a thermometer and obtains the temperature directly, or one can point a radio telescope at the object and derive the temperature by suitably calibrating the receiving equipment. This means that the radio telescope is no more than a giant thermometer used to measure temperatures of objects in the universe.

If we point our radio telescope at each of the planets in turn, we can measure their temperatures, which should only confirm what is already known from infrared measurements, so the two should agree nicely because we do not expect the planets to be anything other than thermal emitters, just like our own planet's surface. Indeed, if the radio astronomer points his telescope down at the ground he will receive a signal equivalent to that produced by any good absorber with a temperature of, say, 300° above absolute zero, which is the equivalent of 27° Centigrade.

It is a well-known law of physics that if an object is a perfect absorber it will also be a perfect radiator; in other words, if it absorbs all the radiation reaching it, it will heat up to a certain point and radiate all wavelengths outward again with an intensity determined by the temperature it has. Such an object is called a *black body,* and it is possible to calculate the way the intensity of the emission from such a black body varies with wavelength, depending on the temperature. Such emission is called *thermal emission,* as mentioned above, and is different from the processes occurring in solar bursts, described above, which are called *nonthermal emission processes.* The radio astronomer also knows that the intensity of the emission produced by various nonthermal processes varies with wavelength in a way that is quite different from the thermal process and he can therefore tell much about the physics of the emitting object merely by studying the way in which the intensity of the received radiation changes as he tunes his radio receiver to different wavelengths. But what has this to do with the planets? It is just this point—that astronomers had no reason to believe that the planets would be anything but simple thermal radiators and instead found a whole different range of phenomena—which we shall now discuss.

Jupiter

Jupiter is well known as a cold and very large planet whose atmosphere consists of frozen methane and ammonia clouds, so it was a considerable shock to radio astronomers to find strong radio signals emanating from this planet. The radio signals indicate the presence of enormous temperatures somewhere on or near Jupiter. Infrared observations had shown its temperature to be at about −120°C, or 150°K (absolute or Kelvin), above absolute zero, and the radio measurements at 3-centimeter wavelength gave approximately the same value. However, measurements at 10 centimeters showed the planet to have an apparent temperature of 600°K; at 21 centimeters it was 2000°K; and at 70 centimeters Jupiter appeared to be at a temperature of 50,000°K. Clearly some other process, quite different from the thermal emission mechanism, was operating. In fact Jupiter was a nonthermal radio source and whatever was producing the radio waves was not on the surface of the planet or even in the atmosphere, both of which were certainly very cold.

At about the time that astronomers were pondering this problem, satellite observations of the intensity of cosmic rays in space had revealed the presence of a series of belts of radiation around the Earth, called the Van Allen belts. These belts contain fast-moving particles trapped by the Earth's magnetic field and held in a doughnut-shaped part of space around the Earth. If Jupiter had such a belt of radiation, and given that Jupiter's magnetic field was much stronger than that of the Earth's, the electrons there might be expected to radiate strong radio waves. This explanation turned out to be the most likely one, since subsequent observations with radio interferometers showed that the source of Jupiter's radio emission consisted of two parts located at several Jupiter diameters from the surface of the planet, as was expected from the picture of Jupiter's Van Allen belts.

Jupiter's Bursts

In addition to the steady nonthermal emission from Jupiter, strong burst-like radiation was accidentally discovered to be originating from that giant planet at wavelengths around 15 meters. This discovery was made by B. Burke and K. Franklin in 1955 during routine radio observations of the sky in the U.S.A. It was part of their program to observe a well-known radio source called Taurus A, better known as the Crab Nebula, every day as a calibrator of their receiver characteristics, but it appeared that a source of interference, in the form of a string of strong pulses, was operative every day shortly after the Crab Nebula had drifted through the region of the sky toward which the telescope was pointed. Closer examination of the records showed that this source of interference was moving slowly relative to the position of the Crab Nebula, and a check of the positions of well-known astronomical objects in the solar system revealed that the radiation was coming from Jupiter.

Examination of other radio records obtained during previous years in Australia revealed that every time the telescope had accidentally been pointed near Jupiter these so-called interference signals had been recorded and everyone had thought that they were just that—interference—and had therefore rejected the data in subsequent analysis procedures. It turned out that there was sufficient "prediscovery" data to enable astronomers to learn quickly some important points concerning the emission of bursts from Jupiter.

The most important property revealed was the periodic nature of the burst activity, which showed a cyclic variation with a characteristic time of 9 hours and 55 minutes. This was not quite the same time as it took the planet to rotate, as determined from observations of the belts and the red spot. These latter observations indicated that the equatorial regions moved somewhat slower than the polar regions, but the burst source was moving at neither of these rates.

Although the emission process for Jupiter's burst (Figure 21) is not yet fully understood it was discovered recently that the characteristic period of 9 hours and 55 minutes was also a characteristic time in the motion of the satellite of Jupiter known as Io, so the source of the radio signals was somehow connected with the motion of that moon of Jupiter. It was not suggested that the radio signals were coming from Io itself but rather that when the line joining Io and the Earth was oriented in some special way to the Earth-Jupiter line, then conditions favored the reception of signals from some part of Jupiter's radiation belts near Io or from some similar region near Jupiter. The actual origin of these bursts is far from being completely understood.

The Greenhouse on Venus

Radio observations of Venus have also posed an interesting question because the surface of that planet is completely covered by clouds. Since the radio signals can penetrate these clouds any signals received from Venus would be expected to be coming from the surface of the planet itself. The clouds are known to be at a temperature of about 230°K (or about −40°C, so it was a considerable surprise when the radio observations showed Venus to have a temperature of 600°K, or about 130°K above that of boiling water. This temperature was found to be constant at all wavelengths in the radio region, which showed that a thermal source had to exist below the clouds with this sort of temperature, but how could this be so?

The answer lies in the presence of the clouds. These appear to act as a sort of greenhouse, which allows radiation from the Sun to seep through to the surface of the planet, where it is then trapped in such a way that the lower atmosphere, and the surface itself, reach a temperature of 600°K. This means that Venus is forever uninhabitable by man and this is the reason that the various Russian space probes that have tried to land there have lasted for such a short time before they stopped sending back signals. Earth-built equipment does not last long in such high temperatures.

The Other Planets

The radio observations of the other planets in the solar system are not as dramatic as the cases of Jupiter and Venus. Mars is a perfectly normal thermal emitter, quite similar to the Moon, which is to be expected from the (now) known similarity of their surfaces. The planets beyond Jupiter also appear to be normal radio emitters, although Saturn might have a weak radiation belt system, similar to Jupiter.

Mercury, however, did turn out to be somewhat unexpected in that it, too, seemed to be too hot when it was first observed. Mercury, the nearest planet to the Sun, was supposed to keep the same side to the Sun always and

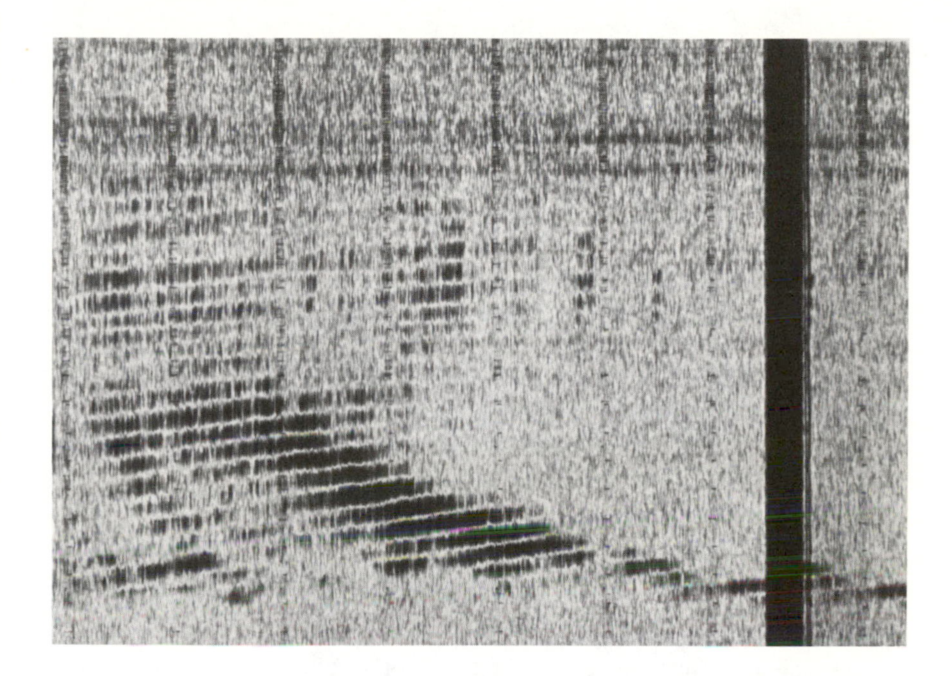

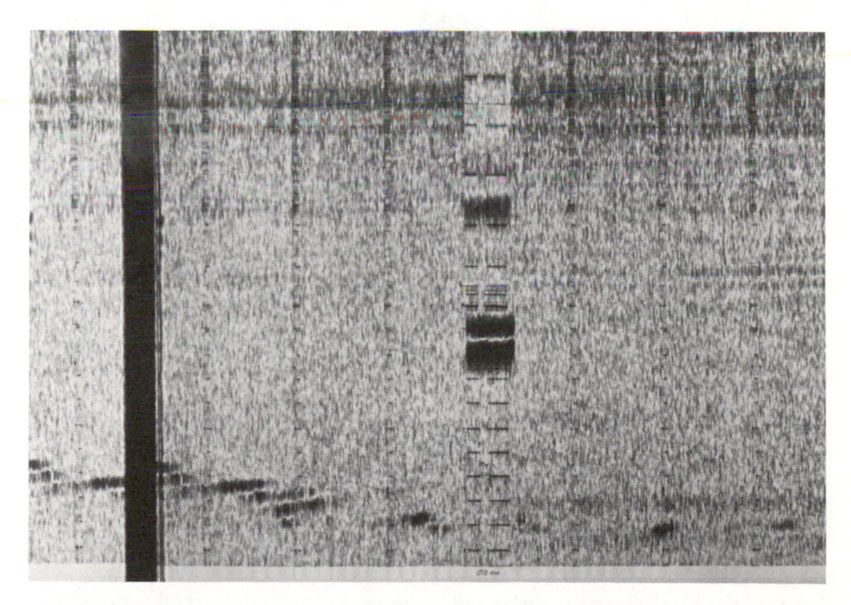

Fig. 21—Two examples of Jupiter radio bursts recorded with the swept frequency received at the radio observatory of the University of Colorado. Vertical lines are calibration markers, and the "snow" on the photographs shows the noise that is produced by the receiver. Time runs horizontally and frequency vertically. (Courtesy University of Colorado, Radio Astronomy Observatory.)

therefore the average temperature of its surface, as derived from radio astronomical observations, should have depended on the amount of the sunlit side visible from the Earth at the time of the observation. This would vary as Mercury moved around the Sun. This so-called phase effect was also expected for Venus, but there it was not seen because of the adequate circulation of the hot atmosphere below the clouds. Mercury was known to have no atmosphere so the phase effect should have been clearly measurable. It turned out that very little phase effect was found and instead the planet always appeared hotter than expected, although it still had the characteristics of a thermal source.

Could there be heating of the planet by radioactivity or some other unknown source of heat such as volcanoes? No. None of these explanations could coincide with the observations. The answer was finally provided by radar astronomy.

Radar Astronomy

The science of bouncing radio signals off other objects in the solar system is known as radar astronomy. Experiments are performed in order to study the way the surfaces reflect radio waves and to learn more about the composition of the planets or to measure the Earth-planet distance by measuring the delay time for the echo. This in turn allows the Sun-Earth distance, called the *astronomical unit* (AU) to be calculated. Without this piece of information no space probes would ever have reached any of the planets. As another result of radar astronomy, the surface features of the moon (Figure 22) and of planets such as Venus (Figure 23) can actually be mapped and the radar observation can also show the way the planet rotates. This latter property is measurable because of the action of the Doppler effect on the returning radio signals. If the radar pulses are sent out at a definite wavelength, the wavelength will be changed a little on reflection, depending on the way the planet is moving toward or away from us. In addition the planet is rotating, which means that one side is coming toward us faster than the other side. This means that the wavelength of the reflected radar pulse will be spread out when it reaches the Earth because some of it will have come from the approaching side of the planet and another fraction will have come from the receding side. The spread in wavelengths of the reflected pulses allows the radar astronomer to state exactly how the planet is rotating.

It turned out that radar observations of Mercury showed that it was not rotating in the way claimed by astronomers for so many years, i.e., once in 88 days with the same face always toward the Sun, but instead it rotated once in

Fig. 22—A radar map of the Moon made using the radar equipment of the 1000-foot radio telescope at Arecibo. (Courtesy N.A.I.C.)

55 days. This meant that Mercury also experiences a day and night, although there the Sun describes a very peculiar path through the sky. The outcome of this is that the night side of the planet is always hotter than it would otherwise be if it had never been subject to the Sun's rays. This explained why the phase effect mentioned above was so weak and why the average temperature of the planet was higher than expected.

Radar observations of Venus have shown that that planet is rotating in a retrograde way, which is opposite to that of all the other planets in the solar system, whose rotation is direct, and is also opposite to that of its orbit around the Sun. It is a peculiarity of the motion of Venus that it always points the same part of its surface toward the Earth whenever it is at its closest approach to the Earth.

By building larger radio telescopes it will be possible, using radar techniques, to accurately map the surfaces of Mars, Venus, and some of the

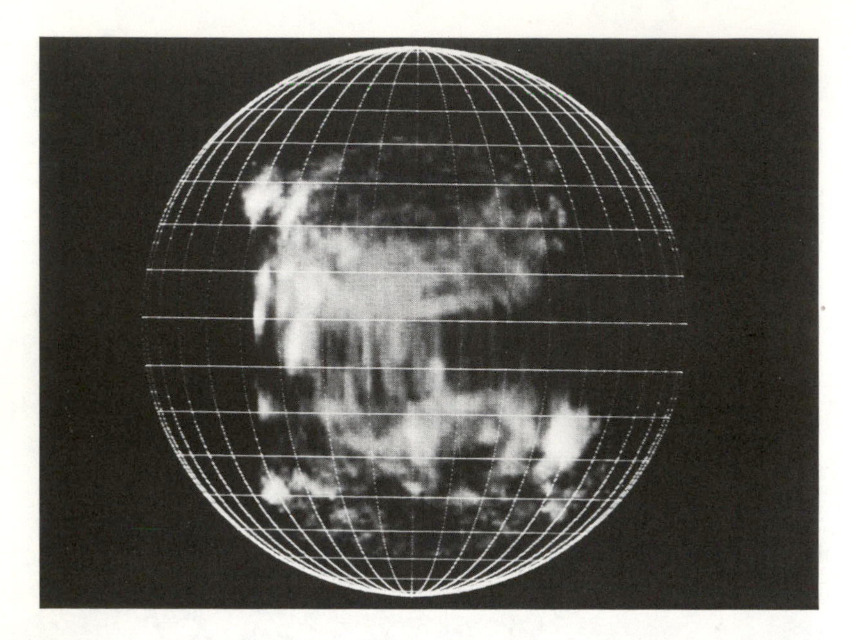

Fig. 23—A map of the radar backscatter from Venus indicating the existence of surface features on the planet, which are obviously not visible to optical astronomers because of the permanent cloud layer covering that planet. (Courtesy M. I. T. Lincoln Laboratory.)

asteroids and the larger satellites of Jupiter. The largest radar transmitting system presently available on Earth is the 1000-foot diameter Arecibo radio telescope in Puerto Rico.

CHAPTER 4

PRODUCING RADIO SIGNALS IN SPACE

Colliding electrons, spiraling electrons, and atomic or molecular transitions are capable of radiating energy in the radio wave part of the spectrum.

Nonthermal or Synchrotron Emission

In the preceding chapter we have mentioned several processes that are operative in the Sun and planets for the generation of radio waves. The one that most often applies to radio signals received from bodies beyond the solar system is the nonthermal radiation process involving emission from fast-moving particles, usually called cosmic ray electrons or relativistic electrons, because of their great speed, which spiral about magnetic field lines in the space between the stars, or inside several types of objects in the universe. This form of emission was originally discovered on Earth in the immediate vicinity of giant accelerators known as synchrotrons. In these machines, particles are accelerated to nearly the speed of light with the aid of strong magnets, and the interaction between the particles and the magnetic fields produced strong radio static in nearby radio sets as well as light emission from the particle beams. It was the Russian astrophysicist I. S. Shklovsky who first stressed that this process might be occurring in space, but this point of view was only slowly accepted at first. It required the definitive measurements of the polarization of the light of the Crab Nebula to prove that this model was probably the best for explaining the mechanism producing the light and radio signals from the Crab, since the polarization found proved that the Crab Nebula contained very strong magnetic fields.

Thermal Emission

Thermal emission is produced by collisions between electrons moving at much slower speeds, determined only by the temperature of the object. These

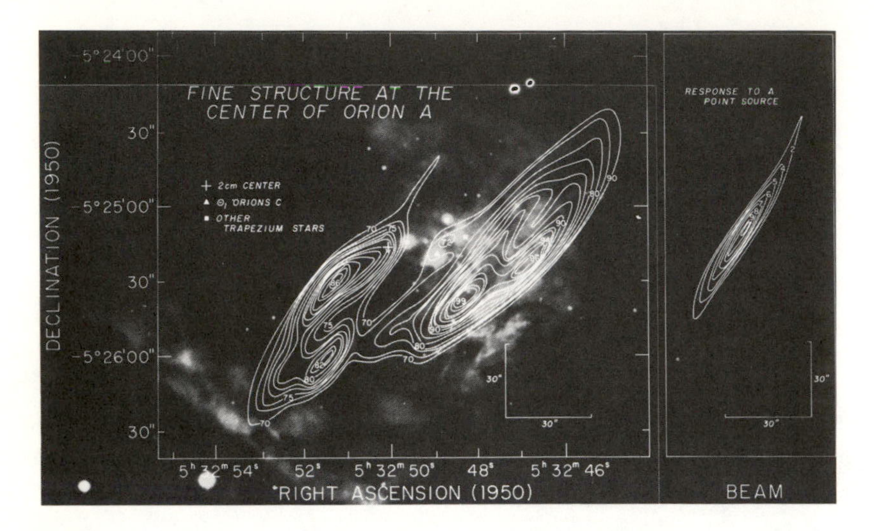

Fig. 24—A radio map of the Orion Nebula, which radiates thermal emission. The apparent elongation of the structure is due to the shape of the telescope beam, also shown. The location of the trapezium stars, the bright stars that heat the nebula, is also shown. (Courtesy NRAO.)

particle motions are called *kinetic motions* and the temperature of the object or cloud of gas reflected by these motions is called the *kinetic temperature.* This concept is somewhat different from our usual picture of temperature as a property measurement with a thermometer. If there are enough particles striking a thermometer we will measure the kinetic temperature, but usually the kinetic temperature is measurement of the speed of motion of the particles concerned, whether or not there are enough around to heat the thermometer. For a thermal radio source (Figure 24) the temperature we measure is the kinetic temperature of the medium, and this will appear the same regardless of the wavelength at which we measure it. This is quite unlike the nonthermal process, which gives a different apparent temperature at different wavelengths, usually being larger at shorter wavelengths.

Plasma Oscillations

Electrons can oscillate in a dense cloud at a rate determined by the density and will radiate signals at a wavelength determined by this rate of oscillation. This process gives rise to some of the radiation from the Sun, and similar emission from distant stars would be just too weak to be picked up on Earth.

Spectral Line Emission

Spectral line emission has not been referred to before, but is extensively referred to in subsequent chapters. Suffice it to say that individual atoms and molecules can emit (or absorb) radio or light signals if their mode of motion changes or if an electron in orbit about the nucleus changes its state of motion. In the case of individual electrons this change can be in the level of its orbit or in its spin about its own axis, and in the case of molecules the change can be in the way the individual atoms are vibrating with respect to one another or in the way the molecule is rotating or the way in which the electrons shared by the constituent atoms might change their orbits or spins.

All these processes have one common property: Whenever an electron changes its state of motion in some way, energy, in the form of electromagnetic waves, may be emitted or absorbed. Indeed, one of these two processes—emission or absorption of radiation—always occurs if the state of motion is in some way changed.

CHAPTER 5

THE MILKY WAY RADIO BEACON

Electrons traveling at the speed of light spiral in gigantic arcs around magnetic fields threading their invisible way through the spaces between the stars. This spiraling generates radio waves and "lights up" the Milky Way band for the radio astronomer.

The Milky Way: Our Galaxy

The bright band of stars visible on a clear night is called the Milky Way, and this name often refers to the galaxy that contains the Sun. Our galaxy is a conglomeration of stars spread throughout a flattened disk-shaped region of space, the diameter being about 100,000 light years. This giant disk rotates slowly about its center once in about 3 billion years, and it is the location of the Sun within this disk that gives the appearance of the Milky Way band of stars that we see at night. One can readily picture this, if one is located within the disk, or wheel, because in certain directions one can see large numbers of stars (for example, toward the center, or hub), and in other directions one can see relatively few stars (directly up and out from the disk).

The Sun is but one of many hundreds of thousands of millions of stars in the Milky Way Galaxy, and there are thousands of millions of galaxies in the universe visible to man.

The center of the Milky Way lies about 30,000 light years away in the direction of the constellation Sagittarius, but is invisible to optical telescopes because of the enormous amounts of obscuring material and stars that lie between it and us. Radio waves, however, can penetrate this dust freely so that radio astronomers can turn their radio telescopes in this direction and observe radio signals originating at or beyond the galactic center.

Radio Signals from the Milky Way

Grote Reber and Karl Jansky first found that the Milky Way was acting like an enormous radio transmitter. But why should it do this? What processes were occurring between the stars that could generate such strong radio signals? We have already alluded to the answer above. It was discovered in the early 1950s that there are magnetic fields threading their way throughout all of the Milky Way. These magnetic fields are now known to be very weak, only about one-hundred-thousandth as strong as the Earth's field and, in addition, man had known for many years that cosmic rays were streaking their way through all of distant space, some of which were even reaching the surface of the Earth. The cosmic rays (high-speed particles) that pass through the Earth's atmosphere and are detected with Geiger counters at the Earth's surface consist mainly of protons and the nuclei of heavier atoms such as helium. From the composition of the particles reaching the Earth's surface, physicists were able to prove that there were also cosmic ray electrons in space that could not penetrate our atmosphere. These have since been detected in satellite experiments, and it is the interaction between these cosmic ray electrons with the magnetic fields far out in space, many light years away, that produce the radio waves from the Milky Way itself. The synchrotron process is therefore occurring in space throughout the galaxy and generating enormous amounts of radio emission.

Cosmic ray electrons are thought to be injected into space as the result of explosions of stars, and they are somehow accelerated to nearly the speed of light by processes not entirely understood. They are forced to spiral about the interstellar magnetic field lines, which, in turn, force them to radiate away some of their energy in the form of radio waves detectable on Earth with radio telescopes tuned to almost any wavelength since the emissions cover an enormous range of wavelengths.

Mapping the Milky Way

Reber had been the first person to map the radio emission from the Milky Way, and other radio astronomers during the last 10 to 20 years have produced many more detailed maps of the distribution of this radiation from outer space. It turns out that because the synchrotron emission process produces relatively more emission at longer wavelengths, the Milky Way appears quite different at different wavelengths. The bright band of increased radio emission associated with the Milky Way band of stars becomes broader and broader as the wavelength of the radio observations is increased, until at very long wavelengths (several meters) one can hardly refer to the Milky Way as a discrete entity at all since the emission is then received equally from all directions in space. This is because the intensity of the received radiation at any wavelength depends in a complex way on the number of electrons involved as well as on their energy. At long wavelengths a depth of only a few hundred light years is enough to produce strong emission, and when we are concerned with regions of space containing large numbers of particles we find that at some point they start to obscure the more distant signals; in other words, they absorb the distant radiation. It is therefore found that the radio signals from the galaxy at 10 meters show this absorption effect, so these parts of the Milky Way now appear less bright to radio astronomers than does the rest of the sky. At this wavelength the Milky Way takes on the appearance of a dark band across the sky! The measurement of the properties of this absorption at the longer wavelengths gives much useful information about conditions in the space between the stars. It turns out that the darkening is produced because the nonthermal emission is absorbed by the thermal electrons in the same part of space. These electrons are nevertheless moving too slowly to produce any observable thermal emission (from interstellar space) themselves.

At shorter wavelengths, such as 20 centimeters or less, the Milky Way band appears narrower and narrower in angular extent until at about 5 centimeters or so it is hardly detected at all.

What are Spurs?

On maps of the Milky Way (Figure 25) one can clearly see tongues of enhanced emission sticking out of the main band of the Milky Way itself. One of these is visible at about 30 to 40 degrees away from the center of the galaxy, where, by the way, the radiation from the Milky Way is greatest, because there the greatest depth of space is contributing to the emission. This tongue is called a *spur*, and the one mentioned above is the most obvious and is called the *North Polar Spur*. The origin of these spurs is far from certain, although their existence has been known for about 20 years. Some believe that they are the remains of very old, and possibly nearby supernovae (100 light years away in the case of the North Polar Spur) and others believe that we see enhanced radiation from these directions because there are many more cosmic ray electrons running out along interstellar magnetic field lines, due to an increased injection of these electrons at some distant point in the Milky Way itself. Such an injection could only be provided by a supernova explosion and differs from the first model in that in the former the spur is part of the shell of matter ejected by the stellar explosion itself whereas in the latter it is not part of a shell at all, but is an illuminated magnetic field line. Let us discuss the exploding stars in some more detail.

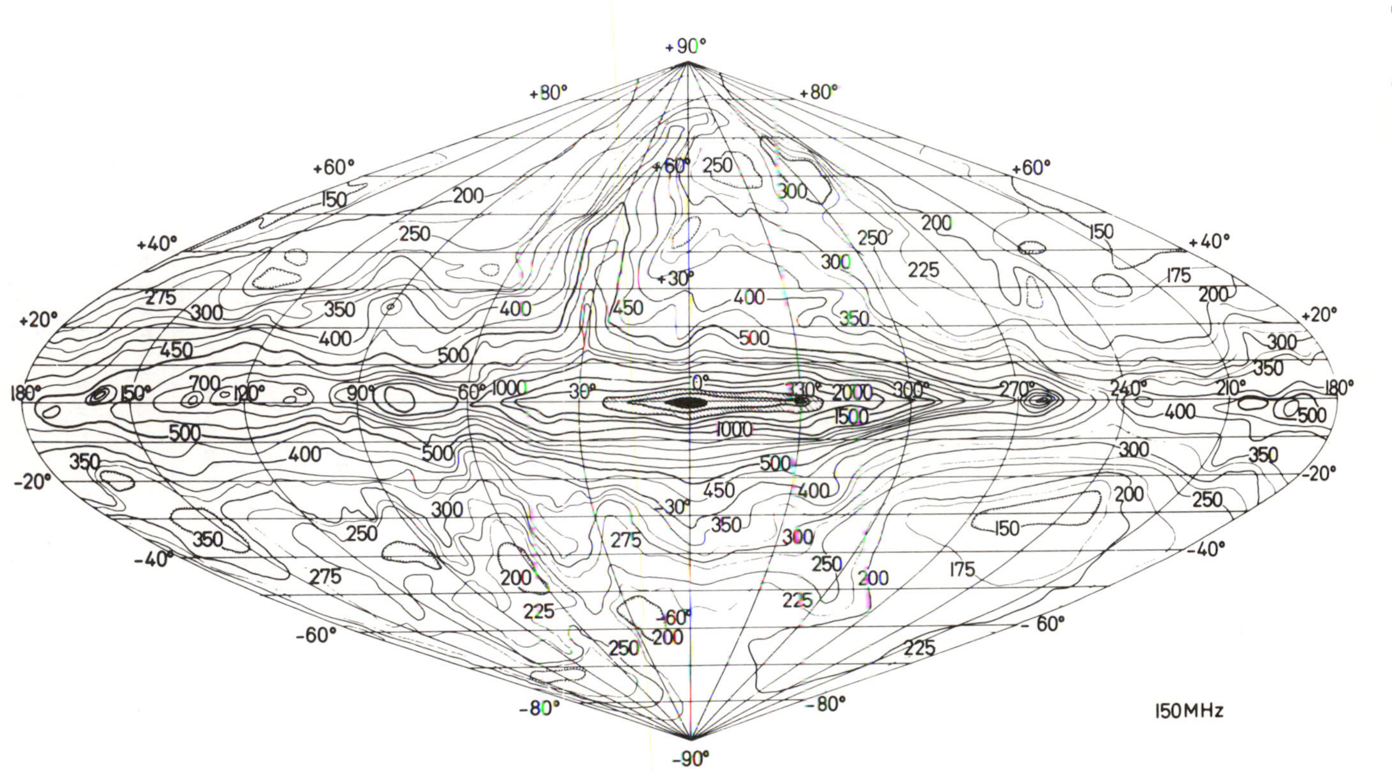

Fig. 25—A radio contour map of the whole sky made at a frequency of 150 MHz (2 meter wavelength). The North Polar Spur is visible at longitude 30 degrees as a projection from the main band of the Milky Way (latitude zero).

CHAPTER 6

EXPLODING STARS

In space are the remains of many stars that exploded violently—some only hundreds of years ago, some thousands of years ago—consuming any planets that might have been orbiting them. These objects are called supernovae and their remains now transmit strong radio waves and are also dramatically visible in photographs.

Guest Stars Make Their Appearance

"On a chi-chou day in the fifth month of the first year of the Chih-ho reign period a guest star appeared at the south east of Thien-Kuan measuring several inches. After more than a year it faded."

So reads one of the old Chinese records which reported the occurrence of an exploding star, or supernova, on July fourth, A.D. 1054. We know that this was the event that gave birth to the beautiful filamentary nebulosity now known as the Crab Nebula (Figure 26). This nebula is the remains of the supernova of 1054, and there are many of these supernova remnants scattered throughout our galaxy.

When a radio astronomer turns his telescope in the direction of this object he picks up strong radio signals because, in addition to the light radiated, the remnant is also a strong source of radio waves. Both the light and radio signals from this object are produced by the synchrotron process, a fact that was proved when the polarization of the light from the nebula was first measured. The synchrotron emission is expected to be polarized at right angles to the direction of the magnetic field in the part of the region at which the emission originates and the photographs showed that this was so. As the polarizing filter attached to the optical telescope was rotated, different parts of the Crab Nebula appeared to shine more strongly, which proved that polarization of

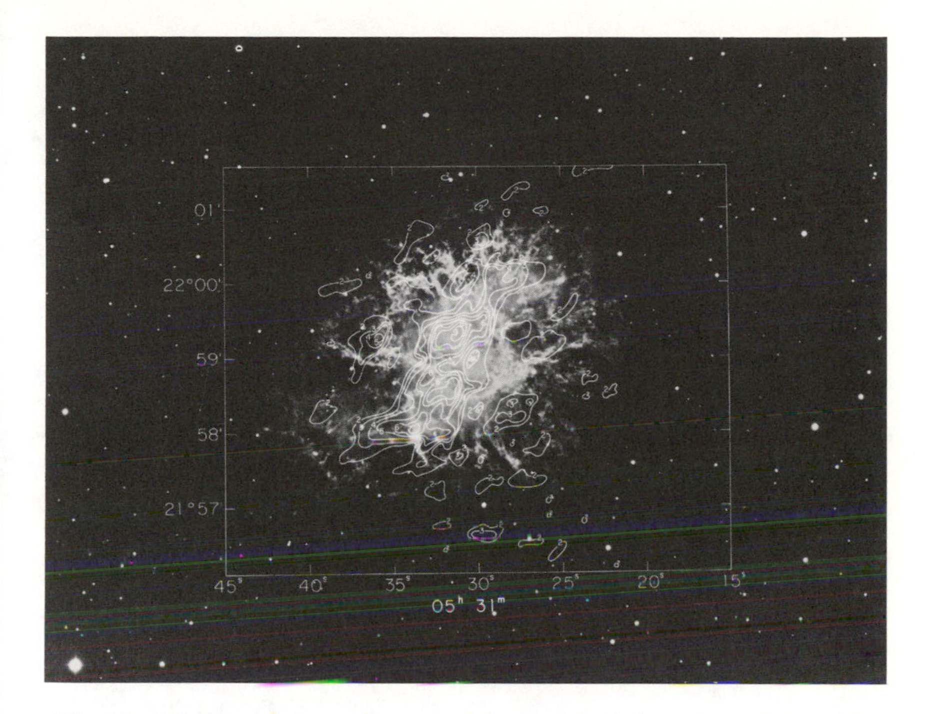

Fig. 26—A high-resolution radio map of the Crab Nebula made at the NRAO at Green Bank with a resolution of 8 seconds of arc. The radio contours are superimposed on an optical photograph of the supernova remnant. (Courtesy NRAO.)

the light was present. If the light were not polarized, then the Crab would have had exactly the same appearance no matter what the angle of the filter had been. A close look at the photographs showed that at any angle of the filter the wisps in the nebula seemed to be elongated at right angles to the filter orientation. This is because the structure of the wisps is apparently controlled by the magnetic fields, which force the particles ejected by the supernova to stream along them. The polarization measured therefore highlighted these phenomena.

In addition to light and radio waves, the Crab also emits strong x rays, which are detected by satellite and balloon-borne equipment. Clearly, if the original star that exploded had had any planets orbiting it at the time, any possible life on them would have immediately ceased to exist there and the planets themselves might have been destroyed.

There are several other rather famous supernovae that were seen to explode in relatively recent times and which are now known to be strong radio emitters. One was seen by Tycho Brahe and another by Johannes Kepler in 1572 and 1604, respectively, but no others have since been seen by the naked-eye observer. If nearby supernova should occur in the near future, they will provide the astronomer with a unique opportunity to study the evolution of such an event from the beginning in all parts of the spectrum; radio, light, x rays, gamma rays, and so forth. Although new supernova are seen at irregular intervals in distant galaxies, they are usually too distant for the radio telescopes presently available to pick up their radio signals.

From observations of the number of supernova remnants in our galaxy and the rate at which they are seen to occur in other galaxies, astronomers estimate that they occur, on average, once in every 60 to 100 years in a galaxy such as our Milky Way system. Clearly we are well overdue for another, since the last one occurred about 350 years ago.

One of the strongest sources of radio emission in the sky, as measured by the strength of the signal it produces in our radio receivers, is the supernova remnant called Cassiopeia A (Figure 27). This object is located at about

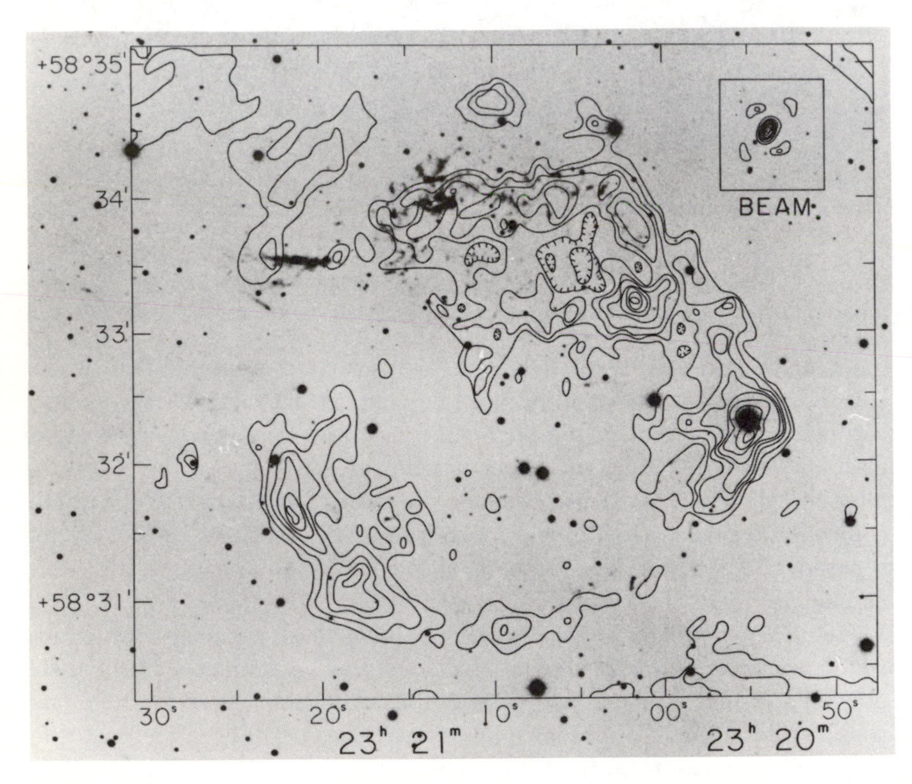

Fig. 27—A radio map of the supernova remnant known as Cassiopeia A superimposed on a negative of the star field. The supernova remnant is heavily obscured by intervening dust clouds, and only some of the filamentary structure is visible as dark blobs in this photograph. The radio map shows a beautiful shell structure with an obvious break where matter has been thrown out to the upper left side. Made with the NRAO interferometer. (Courtesy NRAO.)

10,000 light years from the Sun, and unlike the Crab Nebula, which is about 6000 light years away, the Cassiopeia A source is not readily seen on photographs. That is because the object lies behind some very dense clouds of dust that prevent the light from reaching Earth, although the radio waves pass through this dust completely unhindered. The Cassiopeia supernova has been very carefully studied by radio astronomers and is seen as a ring-shaped object that is still expanding outward at a few percent of its radius per year. Its signals are also becoming weaker at the rate of about one percent per year, and in several hundred years it will no longer appear as one of the strongest radio sources in the sky (see Figure 28).

Fig. 28—A highly specialized radio telescope. This is the horn antenna at the NRAO at Green Bank, West Virginia, which monitors the radio source Cassiopeia A at regular intervals in order to measure how its intensity changes from year to year. Cas A is a relatively young supernova remnant (about 300 years old); it is still expanding and the decrease in its intensity is measurable by means of this radio telescope. Cas A decreases at about 1.5 percent per year. (Courtesy NRAO.)

CHAPTER 7

INTERSTELLAR SPACE: IS IT EMPTY?

Clouds of hydrogen gas drifting through space emit radio signals at the 21-centimeter wavelength and give man direct information on the size and structure of the galaxy and on the temperatures and densities in the hydrogen clouds.

Radio Signals from the Building Blocks of the Universe

The void between the stars is called *interstellar space*, and for many centuries man believed that it was quite empty. Only as recently as 1920 were astronomers able to prove that interstellar space was, in fact, filled with enormous clouds of dust, small particles of unknown composition, which block out starlight. In addition, evidence was found for the presence of calcium and sodium atoms in some of these dust clouds, but it took another 30 years before the most basic element in the universe, the hydrogen atom, was found in interstellar space.

In 1944, Henk van de Hulst, in Holland, suggested that interstellar hydrogen might be detectable because it should emit radio signals at the 21-centimeter wavelength. The suggestion could obviously not be checked immediately, not only because the war was on, but also because radio telescopes and receivers suitable for trying to search for the signals were not available. After the war, however, three groups of radio astronomers started off on a race to pick up the hydrogen signals for the first time. It was virtually a tied race, partly because the Dutch group suffered a damaging fire in their equipment, which set them back by many months. In the early part of 1951 the groups in the U.S.A., Australia, and Holland all succeeded in picking up the radio signals from hydrogen clouds.

Let us consider how the hydrogen atom can transmit radio signals. This atom consists of a nucleus of one proton (positively charged) with a single electron (negatively charged) orbiting around it. It is the hydrogen atom that provides the energy in the Sun by the fusion process; that is, hydrogen is converted to helium. One can liken the hydrogen atom, consisting of the proton with the electron orbiting about it, to the Sun, with the Earth orbiting around it. We know that both the Sun and the Earth also rotate on their own axes, and in just the same way the proton and electron are said to spin on their axes. Here the analogy ends, however, because the electron can, at random times, reverse its spin relative to the proton. If the Earth were to do that, the planet would break up under the stresses that would be produced.

In the hydrogen atom in interstellar clouds, the electron undergoes this rather dramatic reversal of spin every 400 years, and in so doing its state of motion obviously changes considerably. This is equivalent to the atom as a whole gaining or losing energy depending on the precise values of the reversal. Van de Hulst had calculated that this energy difference could, under favorable conditions, be radiated away as a radio signal with a wavelength of 21.2 centimeters. This radiation should be detectable provided that there were enough hydrogen atoms present to give an appreciable amount of radiation and provided that the hydrogen cloud was not at absolute zero temperature. A spin reversal can occur completely spontaneously, and at random, once every 11 million years on average, when one considers only one atom, but van de Hulst realized that if there were enough atoms in a cloud they would collide, inducing spin reversals more often and hence making the radiation more easily detectable. This turned out to be the case, and collisions appear to induce a spin "flip" once every 400 years per atom.

Depending on the way the electron and proton were originally spinning, a radio signal might either be emitted or absorbed by a spin flip. If the two were spinning in the same direction (parallel) and a collision caused them to end up in opposite senses (antiparallel), then a 21.2-centimeter signal would be radiated outward, that is, emitted (see Figure 29). A 21.2-centimeter quantum, as it is usually called, would be absorbed for the reverse case. This means that if a radio astronomer is pointing his radio telescope in the direction of a strong source of radio waves, the Crab Nebula, for example, and if a hydrogen cloud is located between it and the telescope, then the cloud will absorb the radio waves from the Crab at just this 21.2-centimeter wavelength.

Astronomers have always thought that there should be enormous clouds of hydrogen gas in interstellar space, since it is out of such clouds that stars form after contraction to very high densities. It is known that, in some regions of space, star formation is still occurring, and clearly there must be much hydrogen left over for this to still be happening. Before the radio signals at

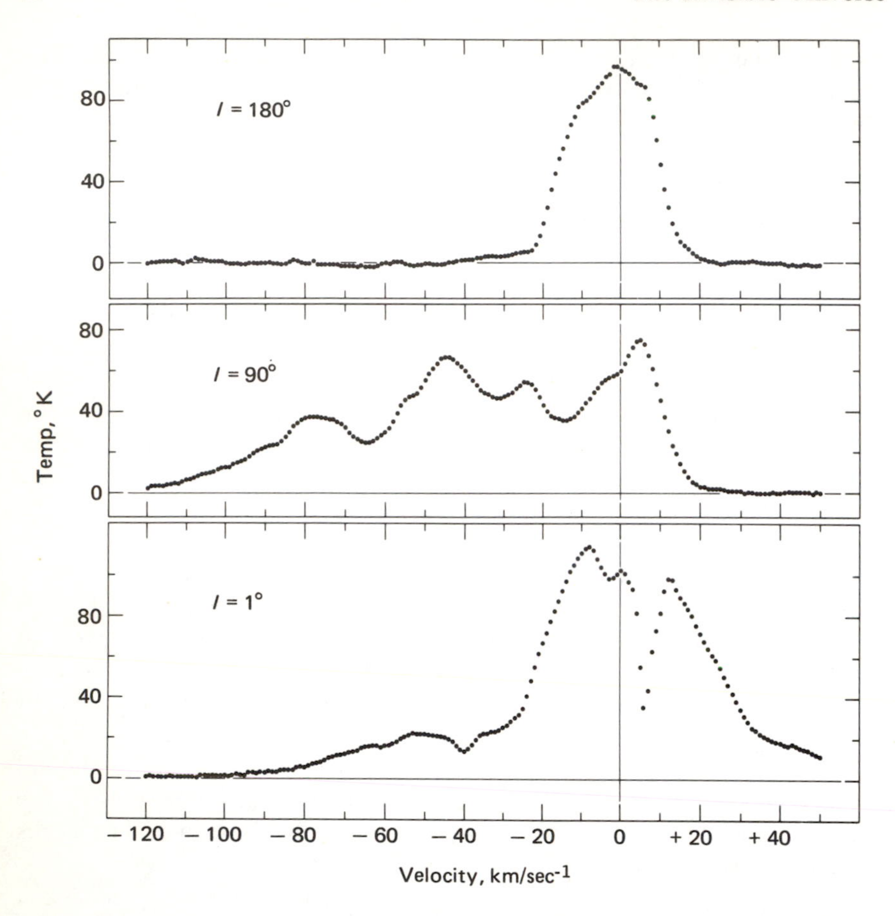

Fig. 29—Examples of 21 centimeter emission spectra from interstellar neutral hydrogen clouds. These three were obtained in three different directions given by the galactic longitude indicated and located at latitude zero. The two scales are temperature in degrees Kelvin and velocity in kilometer per second with respect to the local zero point. Different peaks indicate different distant spiral arms. (Courtesy W. B. Burton.)

21.2 centimeters were discovered, however, there was just no way of studying this fundamental material in interstellar space. The radio observations now give information on the total numbers of atoms in the clouds, the temperature of the clouds, and their distances and sizes.

We shall now discuss just how many atoms are involved in producing the observed radio emission. Bear in mind that the air you breathe contains approximately 10^{19} molecules in every cubic centimeter, but interstellar hydrogen clouds contain only 10 to 100 atoms per cubic centimeter. This is enormously more empty than the best vacuums ever created on Earth. But now consider that a cloud of gas in space might be 10 light years deep, which is equivalent to 3×10^{19} centimeters, so that every square centimeter column through the cloud will contain 10^{20} to 10^{21} atoms. Such a large number of atoms produces a strong enough radio signal to be detected many thousands of light years away. To take this analysis a little further, since every atom is undergoing a transition from one energy state to the other every 400 years and there are about 3×10^7 seconds in a year, we find that in the above cloud, about 10^{10} atoms are emitting energy in any second. Of course half of these are absorbing radiation falling on the cloud from elsewhere, so the true physical picture is somewhat more complicated. However, it is clear that large numbers of hydrogen atoms are involved in the process at any one time.

Taking the Temperature of Hydrogen Clouds

We have already referred to the radio telescope as being a giant thermometer because it can give information about the temperatures in space and this is equally true for the radio observations of hydrogen clouds. The spin-flip transition discussed above occurs as a result of collisions between atoms in the cloud. The hotter the cloud is, the faster the particles move, and the more often they will collide; hence the radio signals emitted by the cloud will be more intense. Therefore it would seem that the strongest signals should be produced by the hottest clouds, but this is not entirely accurate, because when the clouds become too hot the hydrogen atoms are ionized; that is, they lose their electrons. In that case the basic transition giving rise to a 21-centimeter line can no longer take place. In addition, well before this happens, the 21-centimeter line from a hot cloud becomes undetectable for another reason, which is due to line "broadening" produced by the rapid motions occurring in the cloud. The spectral line becomes so broad and the intensity so low that it is very difficult to detect at all.

The Doppler Effect

If an object sending out any sort of wave, whether it be light, radio, or sound wave, is traveling toward or away from us, we will observe a wavelength different from that which was actually emitted. We can picture the case of a motion toward us; the distance between the source and ourselves is decreasing even as the signal is being emitted and, although a certain wavelength was emitted, the waves reaching us arrive more closely together. This is equivalent to a wavelength change in the emitted signal, whether it is a sound tone, a light wave, or a radio wave. The effect is readily observed if we listen to the sound produced by a jet plane as it comes toward us and compare it to the sound we hear as it goes by us and recedes into the distance again.

When the source is coming toward us we have bunching together of the waves, and they appear shorter and the pitch higher, whereas motion away from us produces longer waves and a lower pitch. This effect is called the Doppler effect. A receding motion produces longer wavelengths, which is the equivalent of a *redshift,* because light from a receding object appears redder than it would otherwise, for the same reason.

We mentioned that within a cloud of hydrogen the atoms are moving very fast and continually colliding with one another, and obviously at all times some will be moving toward us and some away from us. This means that the radio signals, which were originally generated at 21.2 centimeters, will have their apparent wavelengths altered by such motions. This, in turn, means that the spectral line is not narrow anymore, but covers a small range of wavelengths determined by the amount of motion in the cloud. This amount of motion is determined by the temperature in the cloud. Radio astronomers are therefore able to convert the observed spectral line widths into a velocity width and hence into a temperature. Typical values for the latter are 20° to 100° above absolute zero in interstellar hydrogen clouds. When the clouds become very hot the line widths become very great, which means that the hottest clouds do not produce the brightest emission lines, but the widest.

The shape of a spectral line in astronomy is determined by the way the velocities of the atoms are distributed. Not all the atoms will have the same velocities; some will be moving momentarily slower because they had collided head on with others, and yet others will be moving faster because they had a collision that reinforced their original motion. The distribution of velocities between the atoms determines the intensity of the line emission, and one observes a distribution which is such that the emission at some average central velocity is greatest and then drops off symmetrically at higher and lower velocities. Such an emission line shape is called a gaussian and is usually observed from clouds in space because of the way energy is shared among the atoms (or molecules) in the clouds. Figure 29 illustrates several complex but more typical, emission line shapes which are the superposition of many gaussians.

But what of the central velocity of the spectral line? Is this always at 21.2 centimeters? No, it is not, because the central velocity is determined by the way the cloud as a whole moves with respect to the observer. It would be 21.2 centimeters if the cloud were at rest with respect to the observer. A cloud coming toward us will appear to emit at a wavelength that is shorter than 21.2 centimeters, usually only some fraction of a centimeter for the motions found in our galaxy, while those clouds moving away will show a redshift and would appear to be emitting at, say, 21.3 centimeters.

Motions toward us are arbitrarily referred to as negative velocities, whereas positive velocities apply to recessional motions.

Looking at the Milky Way from the "Outside" Using the Hydrogen Line

It turns out that the Doppler effect produced by the motion of hydrogen clouds as a whole has very interesting consequences for astronomers because the rotation of the Milky Way results in clouds in different parts of the galaxy having different velocities when viewed from the Sun, depending on their direction and distance from the Sun. This means that observations of the hydrogen emission from all over the galaxy can be translated into the motion of the gas clouds in the line along which the observations were made, and this, in turn, provided we know how fast the galaxy is rotating, can be converted into a distance. This is true because in most directions only matter at a definite distance would appear to be emitting at a particular velocity as a result of the Doppler effect.

Radio astronomers can therefore make maps of how far away the hydrogen in various directions lies, and these data ultimately lead to a picture of what our galaxy looks like, producing a plan view as if we were very far outside the galaxy (Figure 30). This type of analysis is not easy, since the hydrogen is also subject to motions that are not simply the result of rotation around the galactic center. For example, there is a spiral arm in the direction of the galactic center that is moving very rapidly toward us (that is away from the center), probably as the result of some explosion in the central region. Also, there are regions in which the data are difficult to interpret because there are so many clouds present. These clouds may be moving with slightly differing velocities, so that a very complicated spectrum is observed. Lastly, and most importantly, is the fact that the rotational motion of the Milky Way must be accurately known for this type of analysis to be made, and this is very difficult to determine. The problem is that we can determine how the galaxy is rotating only if we know where all the matter is, but we only know where all the gas is if we know how the galaxy is moving!

Information on galactic rotation comes from optical astronomical observations of the Sun relative to the stars around it, but optical astronomers cannot see far enough through the surrounding dust to provide all the answers. They certainly cannot see far enough to obtain the sort of picture produced by the hydrogen line data.

This type of analysis is still being continued in the hopes of obtaining an increasingly accurate picture of what the spiral structure of the Milky Way really is. For example, is our galaxy similar to the Andromeda Nebula (M 31) or is it more similar to M 33 or M 81?

Observations such as these have shown, together with the optical data, that the Milky Way has a radius of about 50,000 light years, with the Sun located about 30,000 light years out from the center (see Figure 30) and

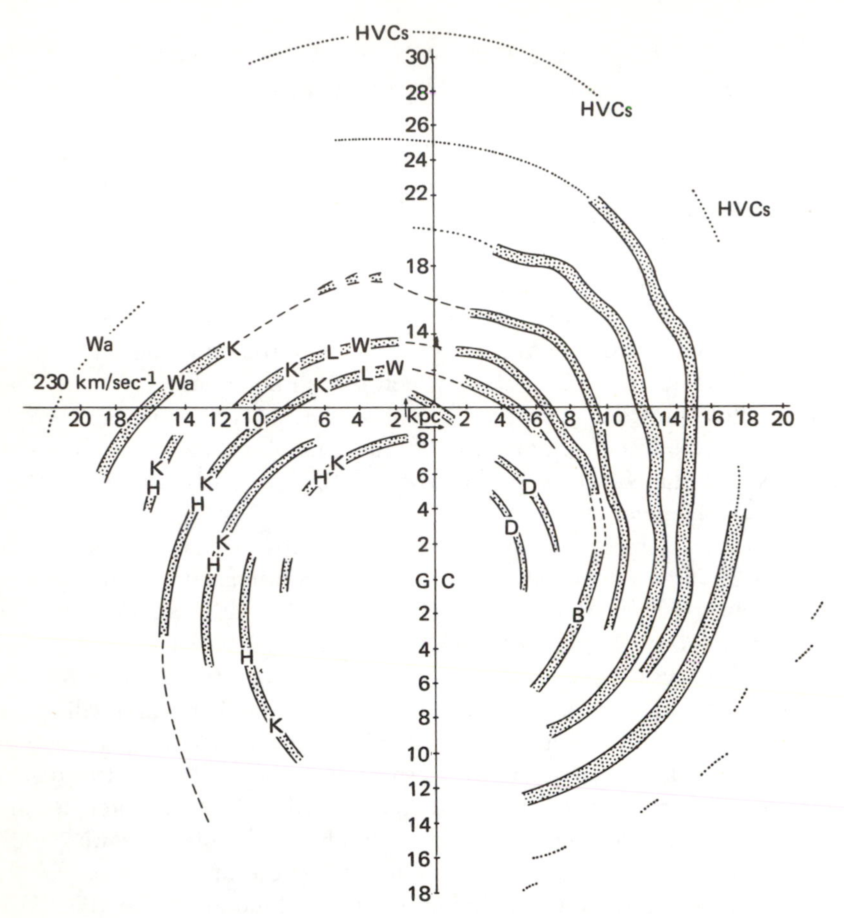

Fig. 30—A recent attempt by the author to make a map of the spiral structure of our galaxy. The Sun is at the origin of the axis and 10 kiloparsecs from the galactic center (GC).

moving with a speed of about 250 km/sec around the center. That is equivalent to about 8 million miles per hour, but since all the stars and gas in the immediate neighborhood are also traveling with this speed, we can never be aware of this motion unless we make the sophisticated astronomical observations necessary to determine it in the first place.

High-Velocity Clouds

One of the strangest phenomena found in the otherwise relatively "normal" field of hydrogen-line astronomy is the presence of large numbers of hydrogen clouds that have large negative velocities in parts of the sky where this is not expected. A negative velocity indicates a motion toward the Sun, and these motions apparently have nothing to do with galactic rotation effects as discussed above, since the matter involved all lies well above the Milky Way plane itself. This suggests that the hydrogen must be fairly local because the disk of our galaxy is supposedly only about 550 to 1000 light years thick, and in astronomy something that close is regarded as local. But why does the matter all appear to be moving toward the Sun?

Several complications exist in working out the answers. Firstly, there are large regions at northern latitudes (referring to the galactic latitude and longitude system determined by the Milky Way band and the direction of the center of the galaxy) over which much of the material has velocities of −50 km/sec, and in addition there are several large clouds of hydrogen apparently traveling at −100 to −150 km/sec when measured with respect to the Sun.

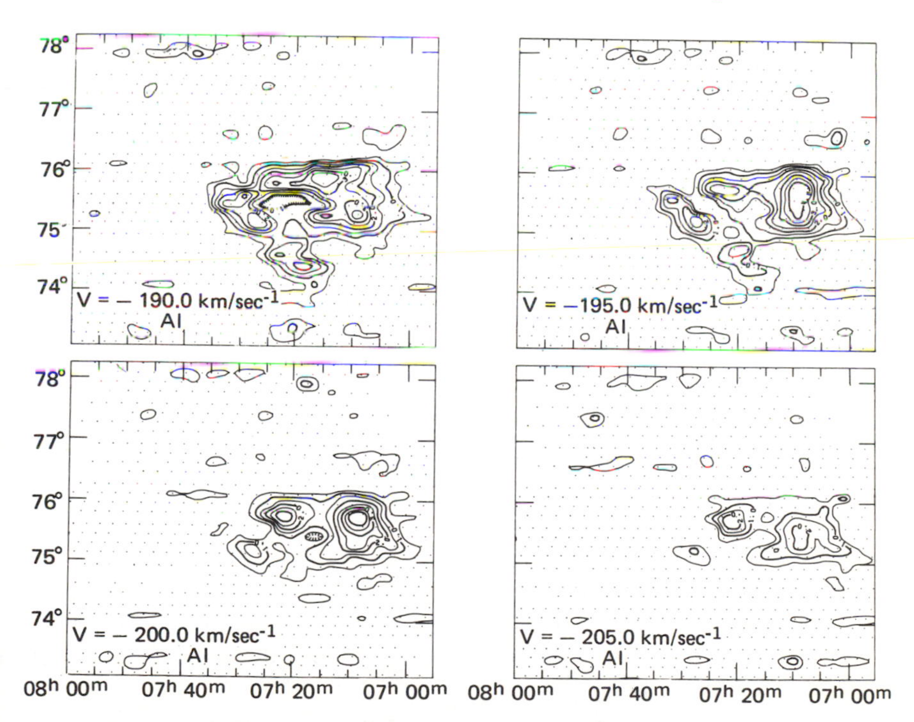

Fig. 31—An example of a set of maps of the emission from a cloud of hydrogen gas as observed at several different closely spaced wavelengths, which are here given in terms of velocities of the Doppler shifts in the cloud. This cloud has a basic velocity of −200 km/sec, which means that it is coming toward us (apparently) at this rate. The coordinates are right ascension and declination. Much structure is found in this cloud, which is one of the so-called high velocity clouds. (Note—a one km/sec change in velocity corresponds to a 4.74 kHz change in frequency of the received signal.)

Furthermore, there are several groups of clouds, some in the direction just about opposite to that of the galactic center and others near the South Galactic Pole, that have velocities of −200 km/sec (see Figure 31). These should be compared with velocities of up to plus or minus 10 km/sec expected for hydrogen clouds moving normally.

No absolutely satisfactory explanation has been put forward to explain all the high-velocity clouds. They clearly have a large amount of kinetic energy in their motion through space if the velocities refer to nearby matter. The only type of phenomenon that could impart such motions to hydrogen clouds is a supernova explosion, but one in which the shell has slowed down to expansion velocities of 50 km/sec, which might be compared with the several thousands of kilometers per second known to exist within visible supernovae such as the Crab Nebula. However, this explanation cannot account for the velocities of 200 km/sec, because in order to impart such motions to hydrogen clouds the matter would cease to be neutral and would become ionized because of the sudden violent accelerations it would be forced to undergo in the supernova shell as it sweeps by the hydrogen cloud.

It is possible that some of the highest velocities are only apparent velocities and might not truly indicate that the hydrogen clouds are moving toward the Sun in local parts of space. For example, we know that the Andromeda Galaxy, which is located about 1.5 million light years away, has an apparent velocity toward the Sun, not because that galaxy is necessarily falling toward us but because the Sun is moving around the center of our galaxy and happens to be moving toward Andromeda at present. If large hydrogen clouds were located at some distance beyond our galaxy, in the direction in which the Sun appears to be moving at present, and if those clouds were not moving toward or away from the center of our galaxy, then from the Sun it would seem that those clouds had a velocity of −250 km/sec. Possibly the clouds with the highest velocities are such clouds in intergalactic space that have not yet had the chance to form either coherent-looking galaxies or stars. The high-velocity clouds do not appear to be associated with any dust or star groups in their direction.

The Dutch radio astronomer Jan Oort feels that the most likely origin for these clouds is the presence of in-falling matter coming into our galaxy from outer space, which is perhaps the last stage in the formation of our galaxy out of the intergalactic medium.

While this book was being written the author proposed that the high-velocity clouds are in fact no more or less than parts of very distant spiral arms that do not have the shape usually associated with spiral arms in the inner part of the Galaxy, i.e., a few thousand light years wide and 600 light years thick. Instead the distant spiral arm is 10,000 to 20,000 light years thick and perhaps equally as wide and at the same time is very patchy (see Figure 32).

Finally, Figure 32 shows a very recent "photograph" of the local, low velocity hydrogen in the sky.

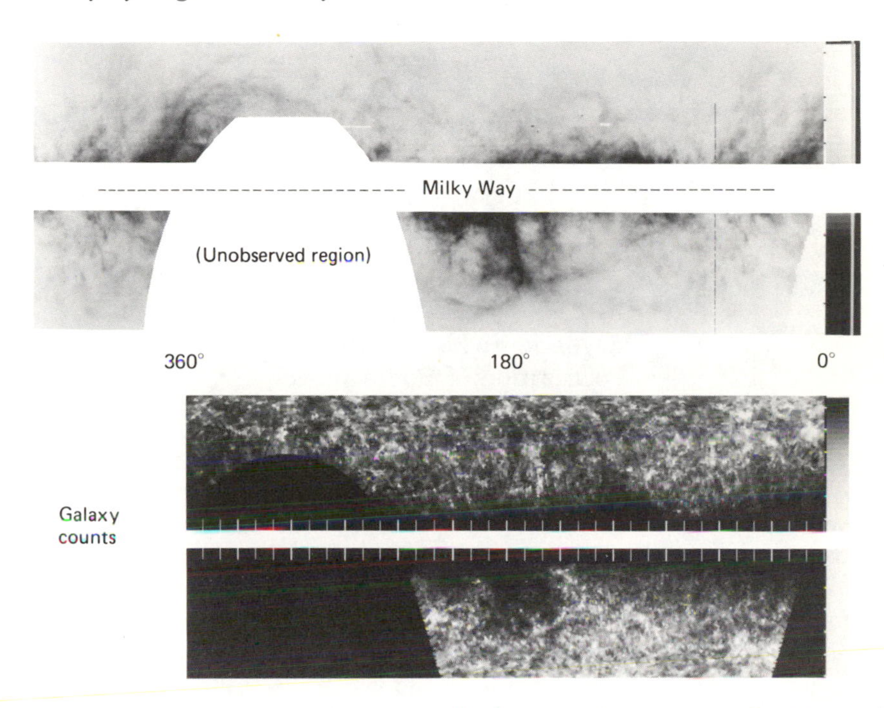

Fig. 32—A recently produced photograph of the distribution on the sky of neutral hydrogen clouds. In the upper diagram the amount of darkening is proportional to the amount of hydrogen gas radiating at the 21-centimeter wavelength. The map covers the full range of galactic longitude of the Milky Way visible in the United States, and the latitude range covered is between plus and minus 60 degrees, with a gap between plus and minus 10, because this area was not included in this survey, and in any case would be totally blackened in this photo. The lower diagram is a similar map representing the galaxy counts over the sky. Dark here should be interpreted as a lack of galaxies visible, i.e., an excess of dust in the Milky Way. (Courtesy C. Heiles and E. Jenkins.)

CHAPTER 8

POLARIZATION IN ASTRONOMY

Magnetic Fields in Space and Polarization of Radio Waves

On several occasions so far we have referred to the presence of magnetic fields in space, in particular when we were discussing the emission of radio waves from the Milky Way itself. But how do we know for sure that magnetic fields exist in interstellar space and how can we determine their strength? The answer to these two questions covers a bit of both radio and optical astronomy, and to consider the first question we must discuss briefly the measurement of the polarization of light from stars. The polarization of any radiation is a measurement of the extent to which the electrical fields set up by the wave as it passes any point are preferentially oriented in any direction. We have pointed out that light and radio waves are two forms of electromagnetic radiation, which means that we can detect the radiation if we use a detecting device sensitive to electric or magnetic fields. In the case of radio waves it is easiest to use an electrical conductor in which electrical currents will be set up when the wave strikes it (see Figure 33). We then feed these currents into the radio set. If the antenna wire is oriented at right angles to the direction from which the radio signal is coming, it will pick up the largest amount of signal. However, one finds in most cases that if, for example, a television antenna is rotated from the horizontal to the vertical, when viewed from the direction of the transmitter, that the signal received decreases or increases. This will happen because the television signal is polarized, which means that the electrical vibrations are occurring preferentially in one direction. If the greatest signal is picked up when the antenna is horizontal, then we can state that the waves are horizontally polarized. If they were 100 percent polarized we would pick up no signal at all in the vertical plane.

And so it is for light waves; they too may be linearly polarized in this way,

Fig. 33—A close-up view of the antenna mount and several of the antennas used with the 1000-foot diameter radio telescope at Arecibo. By moving the mastlike antenna along the support, a point in the sky can be tracked for several hours despite the fact that the reflector itself is fixed to the ground below. The feed platform, as it is called, can also be rotated in the horizontal plane. (Courtesy N.A.I.C.)

although the way to measure linear polarization optically does not involve antennas at all. Instead polarizing filters are used. These are like the polaroid sunglasses that can be easily bought in drugstores. The polarizing filter will allow light with only a particular direction of polarization (the polarization plane of the filter) to pass through it, which means that if one were looking at light that was 100 percent polarized, then one would have all the light transmitted when the filter was aligned parallel to that direction. But rotation through 90° would result in all the light being blocked out. This is because the polarizing filter would stop all polarizations except the one it is designed to pass.

You may readily verify this type of experiment for yourself. You need only two pair of polarized sunglasses. While wearing one, place the other in front of one eye rotating it through 90°. If each glass is perfectly polarizing, no light will pass, and the amount that still passes will give you an idea of how good the polarized sunglasses really are. This is because the first pair allows only one plane of polarization through (it polarizes the light), but the other prevents that plane of polarization from passing through; hence no light at all passes through the two glasses combined.

To illustrate the phenomenon of polarization we can draw a further analogy with a rope being held at both ends but being whipped up and down at one end. The energy generated by this whipping motion travels down the rope as an oscillation and this is completely polarized. Consider, for example, what would happen if a large sheet of plywood with a hole cut in it were placed such that the rope ran through it. If the hole were cut long and vertical, then the vertical oscillations of the rope would be transmitted past this hole with no hindrance, but if the hole were rotated to the horizontal position the vertical vibrations would be completely stopped at the board because the rope would simply not be free to move in that way. We would regard the elongated hole as a polarizing filter, allowing only certain vibrations to get by.

At the same time we notice that the rope can only be made to go up and down at any instant, i.e., move in one plane, and obviously motions cannot occur simultaneously in more than one plane in this case. But for radio waves this is possible. The radiation is then said to be unpolarized, and its analog would be a set of ropes oscillating in all directions simultaneously, which is quite impossible to simulate. It is also impossible for a radio wave produced by one electron to be unpolarized, but the signals found in nature are produced by billions of particles and this usually makes the polarization often appear random and hence unpolarized.

The reasons that various types of radiation are polarized are several. Synchrotron radiation is predominantly linearly polarized because the cosmic ray electrons are spiraling rapidly in helices, or near circles, around the magnetic field lines, and if one looks at these motions from the side, they appear to be almost linear. This is a crude analogy that illustrates the point only approximately. On the other hand, thermal radio emission, produced by billions of electrons rushing about helter skelter and colliding with one another quite randomly show no preferred direction of motion at any instant and therefore the radiation is unpolarized.

There is another form of polarization that we need to know about and that is circular polarization. This can be simulated in our rope experiment by rotating the end of the rope and watching how that wave passes down it. At each point along the rope the motion would appear to be circular and motions could be clockwise or counterclockwise. This is also true for light or radio waves, and we then call the two directions of rotation either right-hand or left-hand polarization. In the case of the synchrotron electrons spiraling in circles, we mentioned that viewed from the side we see the circles nearly as planes, giving linear polarization, but from the front we see circles, which might suggest that we would see circular polarization. This is not so, however, because of the extremely fast motions of the particles involved. It turns out that the cosmic ray electrons radiate only in the direction in which they are traveling at any instant. This means that they radiate out only at right angles to the magnetic field about which they are spiraling and very little energy goes out in the forward or backward direction.

There is a special case of this type of radiation that involves electrons traveling not quite as fast as the speed of light, but which still spiral about magnetic field lines. This also produces radio signals but the process is no longer the synchrotron process and the radiation can be circularly polarized in this case. It is not a common phenomenon in radio astronomy, but is observed to occur in some special events in the Sun.

Now that we have tried to explain a little about polarization let us get back to the question of the magnetic fields in interstellar space.

The Polarization of Starlight

In the previous discussion we referred to the use of polarized sunglasses and the reason that these help to cut out glare is that the light reflected from the ground (the glare) is polarized. The glasses cut out the polarized reflected light and allow the direct light to come through unhindered. Sunlight is originally unpolarized but the reflection of this light off the ground produces linear polarization. This also happens to starlight as it passes through clouds of dust in interstellar space. These dust particles reflect some of the light and allow some through, and in so doing also polarize it. The part we see on Earth is the unreflected part, that is, the part allowed through, which is also polarized. This is because the reflecting, or scattering, particles act as polarizers.

Measurements of the polarization of starlight, when the stars are viewed through intervening dust clouds, show the light to be polarized, which means that the dust particles must have a particular characteristic for this to happen. Measurements of the polarization allow astronomers to make better guesses as to what the dust consists of. The polarization measurements also show that the dust particles must be elongated and aligned in some way for them to produce the amount of linear polarization that they do. The only agent that can be considered as an aligning force is the interstellar magnetic field that controls the orientation of the particles. There is also other evidence for the magnetic fields in the regions of space under consideration. First of all, starlight polarizations suggest that over considerable regions of the Milky Way the magnetic field lines must lie parallel to the plane of the Milky Way and additional confirmation comes from the observations of the polarization of the radio emission from the Milky Way itself. As was mentioned in Chapter 4, this radiation is produced by the synchrotron process, and the theory predicts that this radiation should be polarized at right angles to the magnetic field. Radio observations show that this is so, which not only proves the validity of the synchrotron model for this radiation, but in addition confirms the model needed to account for the starlight polarization, which in these same regions is at right angles to the radio measurements. This is expected because the starlight should be polarized parallel to the magnetic field direction in the dust clouds, whereas the radio polarization is at right angles to the field. All these data tie in very well, and polarization measurements of 7000 stars suggest that the magnetic field within several hundred light years of the Sun is not as simple as was first thought, but rather has the shape of a giant helix.

But what of the strength of the magnetic fields responsible for the polarizations? This is not easy to discover. First of all so little is known about the dust that no good estimate of the field required to orient the particles can be made. In the case of the polarization of the radio emission, one needs to know precisely the energy and numbers of cosmic ray electrons responsible for the emission in order to calculate the field strength required. The best estimates of those properties are based on measurements of the cosmic rays actually reaching the Earth. This is a very uncertain business because a measurement on Earth does not necessarily reflect conditions hundreds of light years away. The best estimates suggest fields of around 10^{-5} gauss. A gauss is the unit of magnetic field, and by comparison the Earth's field is 0.1 gauss and the Sun's field is 1 gauss in the absence of sunspots. There is another way to ascertain the magnetic field strength by radio astronomical methods, however.

Faraday Rotation

A fascinating aspect of the way in which polarized light travels through certain substances was discovered by Michael Faraday. He found that light polarized on entering the crystal in one plane (e.g., vertically) would emerge from the other end with the polarization oriented in some other direction, i.e., it would have rotated about the line in which it was traveling. This same phenomenon occurs for radio waves passing through a cloud of electrons containing a magnetic field. The plane of polarization rotates in passing through the cloud. This process is called the Faraday effect. It results from the fact that a linearly polarized radio wave travels through the cloud as two circular components, right- and left-hand polarized, which travel at slightly different speeds. When they emerge, they recombine to produce a linearly polarized signal. To be more specific we say that their phase velocities are different, causing a phase shift of one wave with respect to the other. Since the angle of the polarized wave in the first place is given by the difference in the phases of the two circular components, we find that a different angle is observed after passing through the cloud since the phases are different. It is possible to measure this effect by making polarization measurements of a given signal with radio telescopes operating at different wavelengths. The amount of Faraday rotation observed depends on wavelength, and thus making observations at many wavelengths allows the radio astronomer to calculate how much rotation has occurred without having to travel out into space through the cloud.

The outcome of this is that measurement of the Faraday effect on a distant object gives information on the magnetic fields as well as on the electron densities in clouds in interstellar space.

CHAPTER 9

MOLECULES BETWEEN THE STARS

Ammonia, water, formaldehyde, cyanide, alcohol, and many other more complex molecules reveal their presence in space by the radio waves they transmit. Now radio astronomers are asking what relevance these discoveries —all made during the last three years—have on our understanding of how life formed on our planet and how common life may be in the universe.

Line Emission from Atoms and Molecules

The emission process described for the case of the hydrogen atom above gives rise to radio signals at a very definite wavelength, and this type of radiation is called spectral line emission. We have already talked about the hydrogen line at 21 centimeters. Other atoms can undergo similar, as well as somewhat different, transitions that might produce spectral lines at different wavelengths. For example, both calcium and sodium show optical spectral lines, which means that they can be observed in interstellar space by the absorption lines they produce in the light from distant stars that passes through clouds of these elements. These two elements produce no radio spectral lines, since the energy differences between the various states of motion of these atoms and their surrounding electrons correspond to radiation in only the visible part of the spectrum. There are several molecules, however, that do radiate radio spectral lines, and the best observed one of these is the combination of an oxygen and a hydrogen atom, commonly called OH or hydroxyl. This molecule undergoes a series of changes in the way it rotates, and the electron, originally associated with the hydrogen atom, is capable of a spin reversal, the net effect of which is that the molecule in its lowest energy state has four possible modes of motion, or four energy levels. Transitions between these levels give rise to four radio spectral lines around the 18-centimeter wavelength, which were first detected in 1963. The fact that all four lines are seen in combination means that the identification of the OH is certain, but let us discuss this point in a little more detail. How do radio astronomers know that the spectral line they pick up is produced by a certain molecule or atom?

Laboratory Checks

The reason that the radio astronomer can be so certain that he is picking up the signature of some particular molecule is that check experiments can be done in the laboratory, which allow physicists or chemists to measure at what wavelength various atoms and molecules absorb radio or light waves. By using the techniques of microwave spectroscopy, they measure precise wavelengths and compile tables of their results, which the astronomer can use when he wants to search for a particular substance or to identify a particular spectral line found. At the same time, mathematical methods for calculating the expected wavelengths of molecules not yet measured in the laboratory also exist. These give an approximate result, which can then be checked experimentally in the lab, or the radio astronomer can tune his receiver to the predicted wavelength and search for that substance in space.

An analogy between these spectral lines and signature tunes is valid because the atom or molecule can be recognized by the tone (that is, the wavelength) of the signals it emits. It is not an audible tone, but a very high pitched radio frequency signal, detectable only with the aid of an expensive radio receiver, either in a controlled experiment in the laboratory or in conjunction with a large radio telescope capable of picking up radio waves from space.

Radio astronomers have already detected the presence of over 20 molecules at the time this chapter was written. Before we discuss the significance of these remarkable discoveries let us relate the story of the detection of the first molecule found by radio astronomers out in the far reaches of space.

The Story of OH and Mysterium

Early in the 1960s, an accurate set of wavelengths at which the OH molecule (or radical as it is more commonly called) would emit or absorb radio waves was produced in the laboratory. Radio astronomers at the Massachusetts Institute of Technology used their 120-foot-diameter dish, which is completely enclosed in a protective radome, together with a newly constructed multichannel receiver, and they tuned it to the predicted wavelengths. The telescope was pointed in the direction of the Cassiopeia A radio source, in whose direction several very dense clouds of hydrogen were known to exist, and it was not long before their observations showed a signal at the exact wavelength predicted for one of the OH spectral lines. There were at least two wavelengths that they needed to examine so as to be absolutely sure that they were dealing with OH, and indeed, after retuning their equipment, the other spectral line was also found.

The OH molecule was expected to show the presence of four spectral lines in the vicinity of the 18-centimeter wavelength, and all four were subsequently identified in the direction of the supernova Cas A and the OH gas was clearly located in the same parts of space as the hydrogen clouds in that direction. This could be ascertained from the similar Doppler shifts found for their emissions. It had also been established that the four OH lines should always occur together with intensities that were fixed relative to one another in the ratio 1:5:9:1. This was predicted theoretically and was shown to be so in the laboratory, and was the result of the way the molecules distributed themselves between the possible energy levels allowed.

In the fall of 1965, I visited several laboratories in the United States and heard rumors that a strong spectral line had been found by radio astronomers studying the OH absorption at an 18-centimeter wavelength. On the East Coast these rumors were very nebulous, but while visiting an important laboratory on the West Coast, I was told, in confidence, that the radio astronomers there had indeed discovered a strong new spectral line at one of the wavelengths expected for the OH emissions, but that this signal was not seen at the other three wavelengths. The discovery had been made during a routine search for new OH clouds located between distant radio sources in the galaxy and the Sun. Theory had suggested that the best chance for finding new OH sources existed in looking at the way the clouds absorbed the radio waves from such well-known radio objects as distant supernova remnants and emission nebulae. This group had, however, found a strong source of spectral line emission instead of an absorption line in the direction of the nebula called IC 1495. Since this spectral line was not seen at the other three wavelengths expected for OH, it was suggested that they had found a new unidentified line, and they proposed that it was produced by some substance, to be determined, which they tentatively called mysterium! There was a precedent for using a name as this in astronomy, since there had been other unidentified lines in astronomy at various times variously called nebulium and coronium.

I promised not to breathe a word about their discovery to anyone until such time as their results were published, and during this time of enforced secrecy I attended a meeting of the American Astronomical Society, where members of several research establishments who had also found the so-called mysterium line in the same source as well as in other directions, were present. Not a word was spoken about these important discoveries at this meeting, however!

Subsequently at least three papers appeared in the scientific literature claiming the first discovery of emission from the OH molecule, for that is what the mysterium line really turned out to be. It still makes fascinating reading to see just who claimed to have discovered what and who ignored the work of others in the first several papers. Such competition still exists between different laboratories and groups of scientists partly because, in the competition for research funds, the previous work done by the individuals is important, and clearly the discovery of something wholly new can be important, not only to the institution involved, but also the individual in his struggle to obtain sufficient research funding or a more secure job.

All these considerations aside, how do we now know that the mysterium signals are in fact OH lines? The answer lies in the fact that more careful measurements subsequently showed this emission to exist in many directions in the galaxy, and often it was seen at all four of the OH wavelengths associated with the OH molecule. The only problem was that the intensity ratios 1:5:9:1 were hardly ever observed. For some reason the OH molecule was radiating much too strongly. Perhaps, some astronomers speculated, an artificial origin was involved. For example, some other civilization, wanting to indicate their presence might use the OH line and transmit it in the wrong intensity ratios so that those picking it up would realize that the lines had an artificial origin. This occurred to me after returning to England from the United States and I asked the group on the West Coast if the signals were also varying in time, which would be expected if the signals were intelligent messages. They soon made more measurements, which showed that the signals were indeed variable, but we now know that both the strong emission and the variability have a natural explanation. The OH signals have become amplified in their passage through space by a process well known to physicists working in the laboratory.

Maser Amplification in Space

Soon afterward, theoretical astrophysicists suggested that the signals picked up were not from any unknown substance, but from OH, and the reason they were such strong signals was that they had been amplified in the clouds in which they arose. It was almost as if something in space were acting as a giant radio set, picking up the weak OH signal and making it much louder, and then transmitting it to Earth! And it did this very selectively, since the other OH lines were not effected.

The mechanism of this process is now fairly well understood. The OH cloud is acting as a giant maser. A maser is a device, constructed in the laboratory, which amplifies radio signals by a very special process called microwave amplification by stimulated emission of radiation. The laser is a variation of this, being a light amplifier or emitter.

The maser action in OH clouds acts to cause the molecules to preferentially emit radiation at a certain wavelength because the states that the molecules find themselves in are no longer distributed between the various rotation or spin states in the predicted way. For some reason the spins of the electrons (for example) might well be preferentially one way so that only emission at one wavelength could occur. A full description of this phenomenon is quite complicated and not fully explained yet, but we have mentioned that the OH has four energy levels, and four transitions between these are allowed. Usually there is a fixed probability that any molecule finds itself randomly in one of the levels. But for some reason most molecules find themselves in a particular set of levels so that only one or two of the transitions are allowed. This anomalous "population" of levels is produced by outside radiation, probably infrared radiation from a nearby star which causes the molecules to first change energy levels drastically to very high levels and then drop down to populate the "normal" levels in an unexpected way.

The study of OH maser sources has shown them to vary in brightness from week to week, and in the direction of some objects, such as the emission nebula IC 1495, the OH sources are incredibly small, with a linear size much smaller than our solar system, and yet they appear as very strong radio emitters. The strongest OH spectral line gives a signal twice as strong as the strongest hydrogen line, yet OH is only one-ten-millionth as abundant as hydrogen. The maser in space certainly amplifies the signals greatly.

Water, Ammonia, and Embalming Fluid

Only in 1968 did radio astronomers find evidence for other molecules in space, and these new discoveries started a wild race to find more, and in particular more exotic, molecules in space. A group in Berkeley, California, under Charles Townes, one of the inventors of the maser, for which he received the Nobel prize, developed a very short wavelength radio receiver and a small telescope (Figure 34) capable of picking up radio signals around 1 centimeter in wavelength. Here both water (H_2O) and ammonia (NH_3) were known to absorb (and therefore could emit) radio waves. There is a table produced by the National Bureau of Standards that lists the wavelengths of thousands of molecules and atoms, but radio astronomers had never seriously considered that any molecules might exist in space. We still do not know why they do exist, but that is getting ahead of our story.

Townes and his group pointed their telescope at the center of our galaxy and discovered both water (Figure 35) and ammonia (Figure 36) in

Fig. 34—The 20-foot-diameter radio telescope of the University of California used for the first detection of interstellar ammonia and water. (Courtesy Radio Astronomy Laboratory, University of California at Berkeley.)

well-known OH and hydrogen clouds in that direction. The correlation is established by the position and the agreement in the velocity shift away from the known zero value for each molecule. This discovery shook the astronomical community considerably, and soon the National Radio Astronomy Observatory was flooded by requests to use the equipment there to search for other molecules whose frequencies were readily available in the National Bureau of Standards tables!

The next molecule discovered was embalming fluid, or more correctly, formaldehyde, which has the chemical formula H_2CO. Its wavelength around 6 centimeters lay nicely in an often-used radio astronomy band protected by international law.

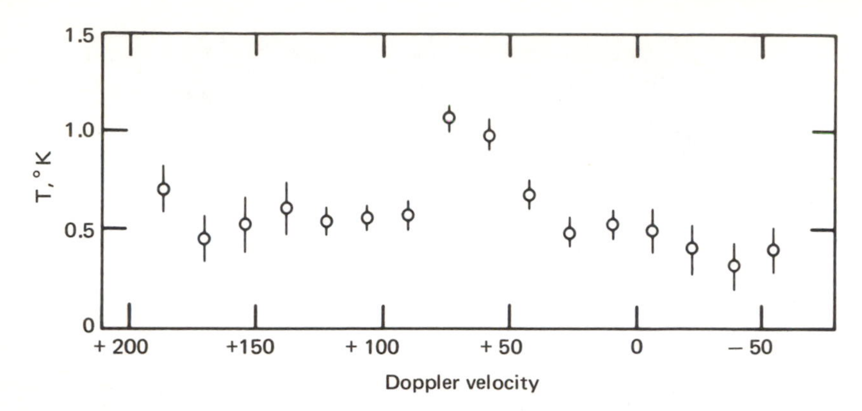

Fig. 35—The first data obtained on the existence of water in interstellar space (December 1968). This observation of the water spectral line was obtained in the direction of the radio source Sagittarius B2, which is now known to contain many different species of molecules. (Courtesy University of California, Radio Astonomy Laboratory.) The increase in level around +60 and +70 km/sec is due to the radio signals from the water molecules in the cloud being observed.

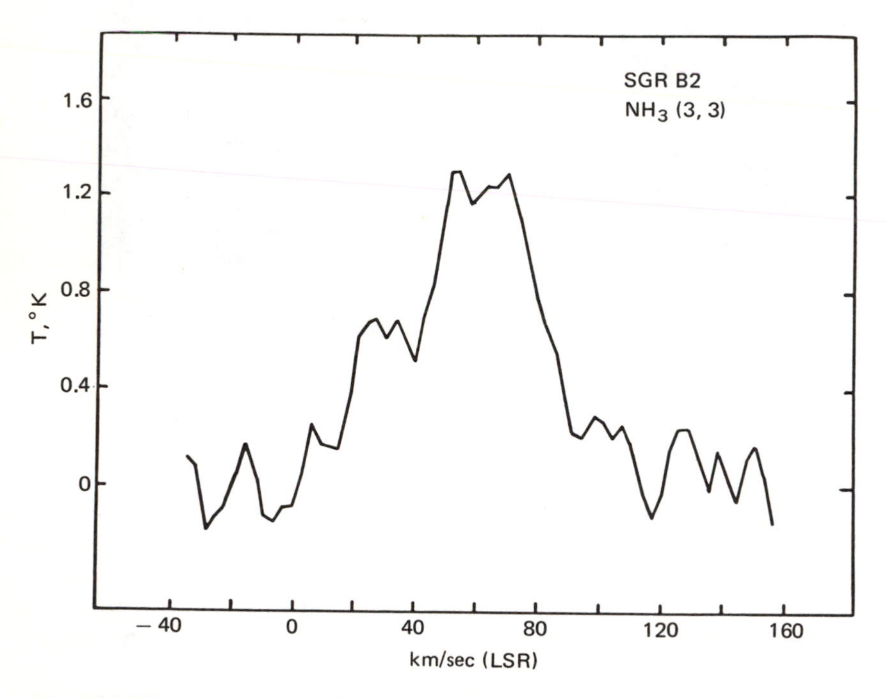

Fig. 36—The emission spectrum of interstellar ammonia at around the 1-centimeter wavelength taken at the Hat Creek Observatory of the University of California. This cloud is the same one in which interstellar water was first found, i.e., the Sagittarius B2 source. (Courtesy University of California Radio Astronomy Laboratory.)

Formaldehyde was discovered early in 1969 and has since turned out to be as widespread in the Milky Way as hydrogen and OH. One of the more dramatic aspects of the discovery of formaldehyde concerns its detection in a dust cloud. Formaldehyde was also widely observed in absorption against the many radio sources in the Milky Way such as emission nebulae or supernovae, but one night the 140-foot telescope was pointed at a famous dust cloud in Taurus. This dust cloud had been found to emit "normal" OH signals with no "masering" action, so the search for formaldehyde emission was made, starting at the dust cloud. There is no radio source behind this dust cloud, but within a quarter of an hour of starting the observation the radio astronomers realized that the signals they were picking up were negative, which meant that absorption had been found rather than the positive-going emission signal expected.

How could formaldehyde absorb radio signals from behind the cloud if there was no source of radio waves to be absorbed? There was only one answer: The cloud was absorbing the radiation coming to us as a leftover of the "Big Bang" by which the universe started, according to some cosmologists.

This background radiation is all around us no matter where we look and appears as a 3° (above absolute zero) radio signal to the radio astronomer. In order to absorb this background signal the dust cloud had to be colder than 3°. However, that is not possible because just as any object sitting inside an oven will finally reach the temperature of the oven, so any cloud in the universe will heat up to 3°.

This dilemma is far from solved, but loosely one can say that an inverse maser operates. Instead of radio emission being enhanced by the way the energy levels of the molecule are populated, they are, in the case of formaldehyde, weakened. No one knows precisely why as yet.

At the same time we should note that the water signals from interstellar space are also amplified by a maser action and the water emission line (Figure 37) is 100 times stronger even than the OH line (Figure 38). It would have been possible to discover interstellar water clouds in the 1950s if someone had thought to look!

Other Interstellar Pollutants

Ammonia and formaldehyde are not the most gentle of substances, but since their discovery several other very toxic, and also very important, gases have been found in the so-called empty voids of interstellar space.

The list includes cyanogen (CN), carbon monoxide (CO), hydrogen

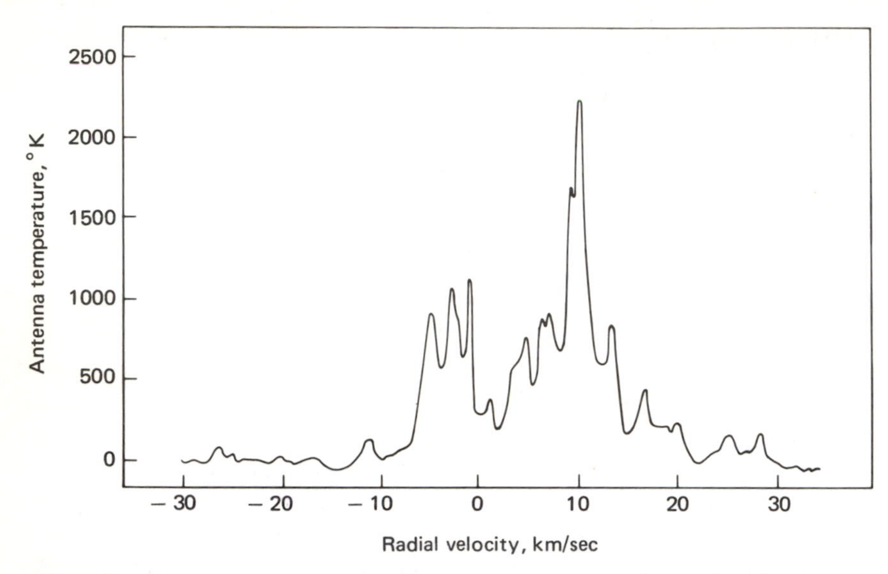

Fig. 37—The very complex spectral lines due to water as observed in the direction of the emission nebula W49 with the 140 foot radio telescope at Green Bank. These lines are undergoing maser amplification in space, which accounts for their great intensity. The radio source W49 is invisible optically, being totally obscured by dust. (Courtesy NRAO.)

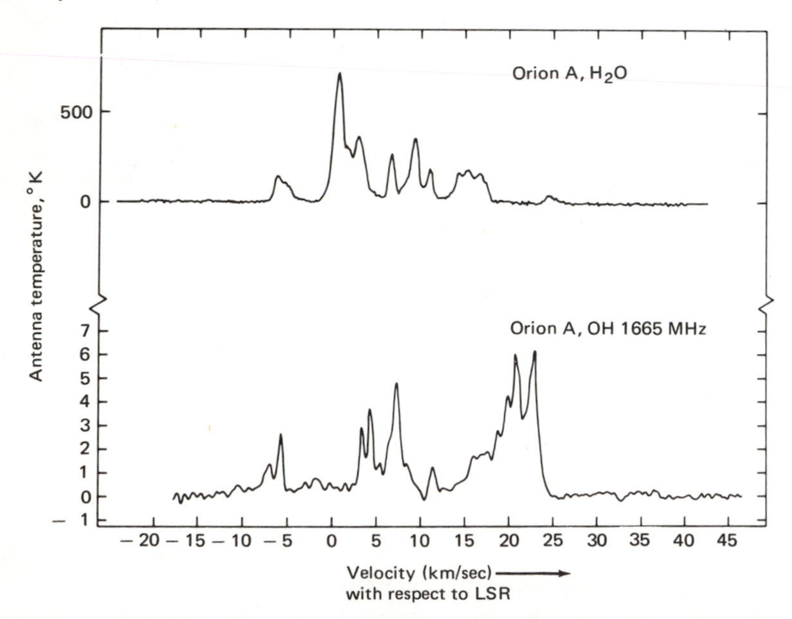

Fig. 38—Comparison of the 1.35-centimeter water emission line and the 18-centimeter OH emission line observed in the direction of the Orion Nebula (Orion A). Note the similarities and differences. Both molecules are masering. (Courtesy NRAO.)

cyanide (HCN), cyanoacetylene (CH_3N), methyl alcohol (CH_2OH), formic acid (or ant juice) (HCOOH), and carbon monosulphide (CS). Methyl alcohol is not the type of alcohol we drink (which is ethyl alcohol), so man need not think about sending a spaceship out there to collect it! Even if it is found, the number of molecules of any of these species per cubic centimeter is so small that one could never collect enough to detect it chemically, let alone use it!

The Bandwagon Effect

Molecules in space are clearly exciting discoveries and in the modern world of research, which is not always the gentle occupation it seems, the survival of many depends on the work they do and the papers they publish. And so it is in radio astronomy. To be able to have one's name associated with the discovery of a new molecule is thought by many to enhance their reputation and so make more secure one's future. Clearly an experiment to detect a new molecule is worth a paper or two in the scientific journals, and since one's tenure in the U.S. university system depends on papers published, many researchers, not otherwise clearly motivated, jumped onto the molecule bandwagon, and many have succeeded quite well.

Together with the rat race to discover a new molecule went the "secrecy syndrome," in which the results of one's experiment were not even whispered to anyone in case another competitor should hear about it and publish a similar result first. This phenomenon was manifested in 1965 with the mysterium discovery, and similar situations since then have hampered good communication between astronomers. The situation does not appear to have gotten out of hand yet, but at one time in 1969 I was a little surprised to hear about the formaldehyde discovery, made by colleagues at the NRAO, two of whom had offices only yards away from mine, from a colleague in California who had heard it from someone at M.I.T. in Cambridge, Massachusetts, who had heard it from someone . . . etc.!

Astrochemistry

Together with these dramatic discoveries of molecules in space comes the birth of astrochemistry, which is the study of chemical processes in space. Knowing how these molecules are formed in space is clearly important to man, since many of the molecules found are basic to prebiotic chemistry; that is, the stages just before life formed on Earth. We shall discuss some of this in Chapter 19 on life in space, but at present suffice it to say that astrochemistry is very young and could become one of the most important, if not the most important, branch of all astronomy. In 10 months from 1970 to 1971, five conferences were organized with the express intention of bringing together chemists and astronomers to discuss these problems. This reflected the rapid development in the field.

Table 1 is a list of spectral lines observed by radio astronomers.

Table 1

Spectral Lines in Radio Astronomy

				Radio telescope	
Year	Molecule or atom	Symbol	Wavelength	Location	Size
1951	Hydrogen (neutral)	HI	21.1 cm	Harvard Obs.	Horn
1963	Hydroxyl	OH	18.0 cm	Lincoln Labs.	84 ft
1964	Hydrogen (ionized)	HII	3.4 cm	Lebedev	22 m
1966	Helium (ionized)	He	18.0 cm	Harvard	60 ft
1967	Carbon (ionized)	C	6.0 cm	NRAO	140 ft
1968	Ammonia	NH_3	1.3 cm	Hat Creek	20 ft
1968	Water	H_2O	1.3 cm	Hat Creek	20 ft
1969	Formaldehyde	H_2CO	6.2 cm	NRAO	140 ft
1970	Carbon monoxide	CO	2.6 mm	NRAO	36 ft
1970	Cyanogen	CN	2.6 mm	NRAO	36 ft
1970	Hydrogen cyanide	HCN	3.4 mm	NRAO	36 ft
1970	X-ogen	?	3.4 mm	NRAO	36 ft
1970	Cyanoacetylene	HC_3N	3.3 cm	NRAO	140 ft
1970	Methyl alcohol	CH_3OH	36.0 cm	NRAO	140 ft
1970	Formic acid	CHOOH	18.0 cm	NRAO	140 ft
1971	Carbon monosulfide	CS	2.0 mm	NRAO	36 ft
1971	Formamide	NH_2COH	6.5 cm	NRAO	140 ft
1971	Carbonyl sulfide	OCS	2.7 mm	NRAO	36 ft
1971	Silicon monoxide	SiO	2.7 mm	NRAO	36 ft
1971	Methyl cyanide	CH_3CN	2.7 mm	NRAO	36 ft
1971	Isocyanic acid	HNCO	3.4 mm	NRAO	36 ft
1971	Hydrogen isocyanide	HNC	3.3 mm	NRAO	36 ft
1971	Methyl acetylene	CH_3CCH	3.5 mm	NRAO	36 ft
1971	Acetaldehyde	CH_3CHO	20.0 cm	NRAO	140 ft
1972	Thioformaldehyde	H_2CS	9.5 cm	Parkes	210 ft
1972	Hydrogen sulfide	H_2S	1.8 mm	NRAO	36 ft
1972	Methanimine	CH_2NH	5.7 cm	Parkes	210 ft
1973	Sulfur monoxide	SO	2.9 mm	NRAO	36 ft

CHAPTER 10

LOCATING RADIO SOURCES IN THE UNIVERSE

Millions of radio emitters that are not stars in our own galaxy probably exist in the universe. Only a few thousand of these "radio sources," as they are called, have been discovered, accurately cataloged and located in space. Many of the radio sources coincide in position with visible objects such as supernovae or galaxies or quasars, but many of them show no optically identifiable counterpart. Some of these may well be objects at the very edge of the universe, beyond the vision of the largest optical telescopes.

What Is a Radio Source?

A radio source refers to an object somewhere in space that emits radio signals that we can detect on Earth using radio telescopes. The term "radio star" was often used in the early days of radio astronomy, but when it was discovered that no radio emitter could positively be associated with any known star (apart from the Sun), the name *radio source* was used instead. Recently several true stars have been found to be radio emitters and we shall discuss them separately later.

Locating the Position of Radio Sources

Radio telescopes are used to map the skies, which means that they produce a record of the way the intensity of the received radio signals vary over the sky. When a distinct increase in level is noted, which is clearly greater than the average noise level on the record, the deflection is usually attributed to another radio source. The position in which the telescope was pointing at the time is simply noted and the source given an identifying number.

One of the earliest major surveys was the third Cambridge (England) survey of the sky, and the radio sources they located are called 3C1, 3C2, etc., to 3C464, in rough order of position (see Figure 39). Other more easily recognizable numbering systems have subsequently been adopted by other observatories, notable Parkes in Australia, where the sources are called PKS, followed by a number indicating their approximate position in right ascension and declination.

Knowing the exact position of the sources allows the radio astronomers to refer to optical photographs of the sky, such as the giant Palomar Atlas of the sky, to try to identify which visible object might be giving out the strong radio signals (see Figure 40). In the early days of radio astronomy the more obvious identification with dramatic phenomena such as the Crab Nebula were relatively easily made, but nowadays the task is considerably more difficult.

We have noted in an earlier chapter that a single radio telescope cannot pinpoint the position of a radio source to better than a minute of arc or so in angle, depending on the size of the dish, but over such a region of the sky there will be many objects visible to the world's largest optical telescopes. Clearly one needs more than a single dish for identification purposes. So an

Fig. 39—Part of the interferometer of Cambridge University in England used for making general surveys of the sky in searches for new radio sources as well as for obtaining counts of the numbers of radio sources that might have a bearing on the problems of cosmology. (Courtesy Mullard Radio Astronomy Observatory.)

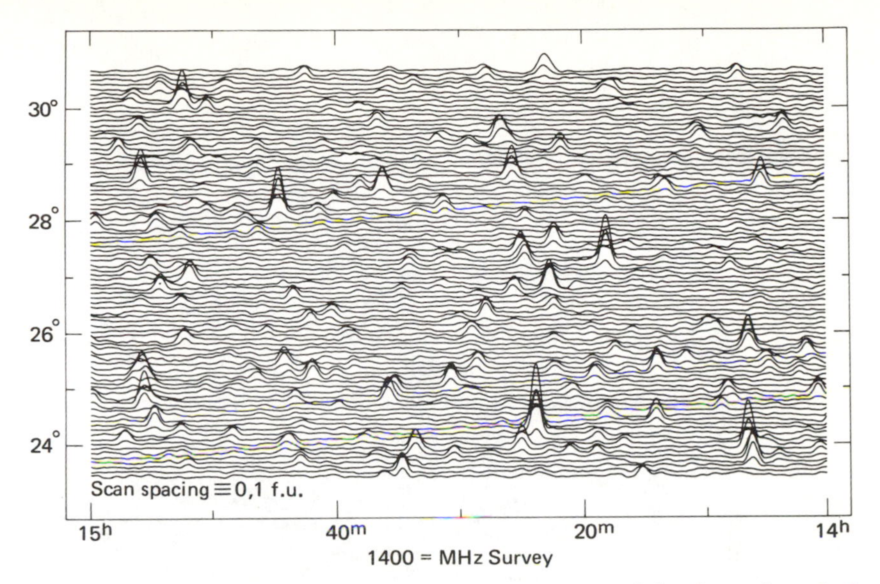

Fig. 40—A very high-sensitivity survey of a small region of the sky made with the 300-foot telescope of the NRAO at Green Bank. This method of data display is different from the contour maps usually produced, but it contains the same information. Each of the peaks is a separate radio source, and if one were to try to identify all the sources at this low level of intensity with the optical objects, he would find that in many cases nothing would be visible on the photographs at the position of many of these weak radio sources. Several would be quasars, perhaps, and many would be very distant galaxies [f.u. = flux units]. (Courtesy NRAO.)

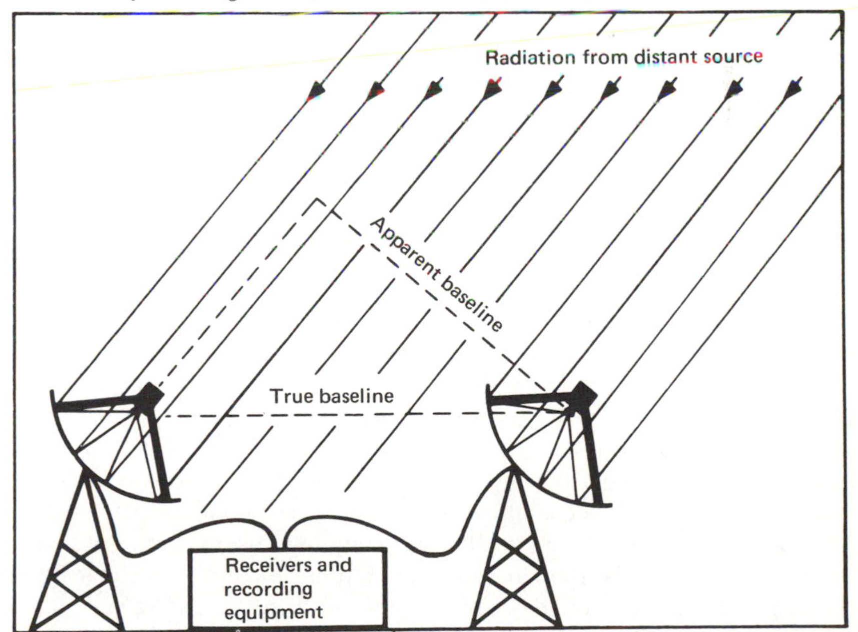

Fig. 41—A schematic indicating the operation of a simple inteferometer. The baseline is effectively the diameter of an equivalent single-dish antenna that would have to be built to obtain the same resolving power. (Courtesy NRAO.)

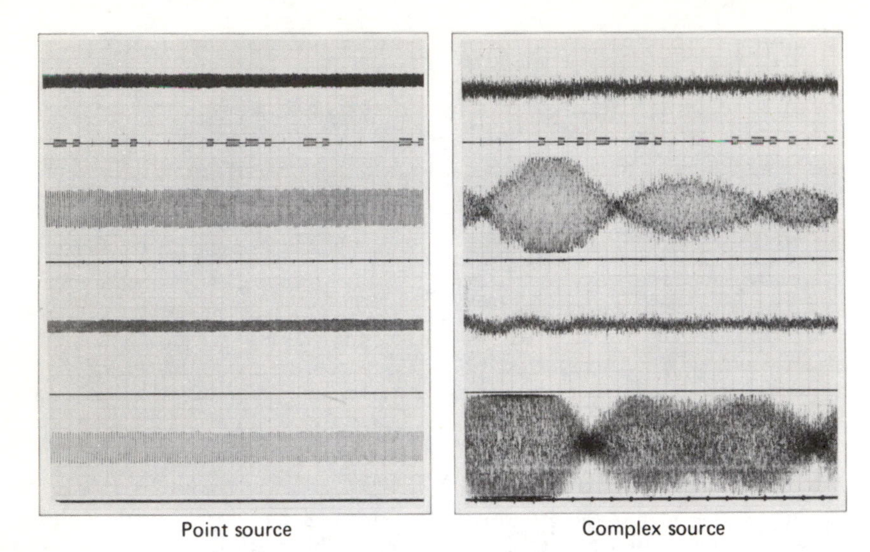

Fig. 42—Examples of the type of fringes produced by an interferometer when it observes various types of radio sources. It is obvious that such records would be difficult to process by hand, so the computer is usually used to record and subsequently examine these types of records. However, records such as these are usually obtained from a chart recorder while the observations are in progress so that the astronomer or the telescope operator can monitor the progress of the experiment.

interferometer is often used. An interferometer consists of two or more radio telescopes joined by electrical cables, and the radio signals detected by all of them are fed to a central receiver where the signals are combined and then displayed (Figure 41). As the radio source moves across the sky, the radio signals from two dishes will successively reinforce and then cancel out one another, producing *interference fringes* (Figure 42). Two telescopes, separated by, say, a mile, will be simulating the ability of a single dish a mile in diameter to pinpoint the position of radio sources to much greater accuracy than is possible using one only 100 feet in diameter. In fact the resolving power, which is the ability to pinpoint the source and see structure within it, depends on the distance between the dishes. The farther apart, the better they can locate small radio sources. Unfortunately there is a limit to the distance at which they can be placed because the farther apart they are the more difficult it is to get the received signals together at a single place. The cables needed for this become too long and their electrical properties do not stay constant for long enough to allow an accurate measurement of position of the radio sources, since the measurements require a stability in the length of the cable or the phase of the signals being received, which depends on the cable length as well.

Despite this drawback, an interferometer with large separation between the dishes, or a long baseline as it is called, can still discriminate very small scale structure in the sources although it might not know exactly where the source is! It might be able to "see" structures of a thousandth of a second arc in size, while not being able to pinpoint the source to 10 seconds of arc in position! As a result long baseline (10 miles to many thousands of miles) interferometers are used primarily for detecting small-scale structures in sources, whereas short baseline (less than a few miles) interferometers are used to locate radio sources accurately.

Table 2 is a list of the positions and radio intensities as well as the identification of some of the brightest radio sources observed in the northern hemisphere.

The Moon as a Shield

A novel way of measuring the position of radio sources is to allow the Moon to eclipse, or as it is usually called, occult, the radio source. Since we know very accurately where the Moon is at any instant, the radio astronomer merely watches a radio source, whose occultation is roughly predicted, until it disappears and later reappears. Measurement of the times of these events allows the calculation of the position of the radio source. Under the right conditions the inaccuracy in the measurements need be only our uncertainty of the shape of the mountains of the moon behind which the source has disappeared. We shall discuss this experiment a little more later on, but let us now discuss one of the most interesting early identifications of radio sources.

The Identification of Cygnus A

In radio astronomy one of the conventions is referring to several of the strongest radio sources in the sky by the constellation in which they occur, followed by A or B, etc., depending on whether they are the strongest, or second strongest, and so on, in that constellation. This nomenclature never caught on, although it has stuck to the sources discovered earliest. The Crab Nebula, for example, is Taurus A and the supernova in Cassiopeia is Cas A.

In Cygnus there is also a very strong radio source, and early examination of the photograph showed no obvious supernova or emission nebula (as for example Orion A, the Orion Nebula) at the position of Cygnus A.

The radio position of Cygnus A was obtained with the Cambridge interferometer in 1951 by Graham Smith and sent to Walter Baade, who

Table 2

The Brightest Radio Sources Visible in the Northern Hemisphere (Based on Observations at the 20-Centimeter Wavelength)

Name	Right Ascension			Declination			Intensity (flux units)	Identification
	hr	min	sec	deg	min	sec		
3C 10	00	22	37	63	51	41	44	Supernova remnant[a]–Tycho's supernova
3C 20	00	40	20	51	47	10	12	Galaxy
3C 33	01	06	13	13	03	28	13	Elliptical Galaxy
3C 48	01	34	50	32	54	20	16	Quasar
3C 58	02	01	52	64	35	17	34	Supernova remnant[a]
3C 84	03	16	30	41	19	52	14	Seyfert Galaxy
Fornax A	03	20	42	−37	25	00	115	Spiral Galaxy
NRAO 1560	04	00	00	51	08	00	26	
NRAO 1650	04	07	08	50	58	00	19	
3C 111	04	15	02	37	54	29	15	
3C 123	04	33	55	29	34	14	47	Galaxy
Pictor A	05	18	18	−45	49	39	66	D Galaxy[b]
3C 139.1	05	19	21	33	25	00	40	Emission nebula[a]
NRAO 2068	05	21	13	−36	30	19	19	N Galaxy[c]
3C 144	05	31	30	21	59	00	875	Supernova remnant[a]–Crab Nebula–Taurus A
3C 145	05	32	51	−05	25	00	520	Emission nebula[a]–Orion A–NGC 1976
3C 147	05	38	44	49	49	42	23	Quasar
3C 147.1	05	39	11	−01	55	42	65	Emission nebula[a]–Orion B–NGC 2024
3C 153.1	06	06	53	20	30	40	29	Emission nebula[a]
3C 161	06	24	43	−05	51	14	19	
3C 196	08	09	59	48	22	07	14	Quasar
3C 218	09	15	41	−11	53	05	43	D Galaxy[b]
3C 270	12	16	50	06	06	09	18	Elliptical Galaxy
3C 273	12	26	33	02	19	42	46	Quasar
3C 274	12	28	18	12	40	02	198	Elliptical Galaxy–M87–Virgo A
3C 279	12	53	36	−5	31	08	11	Quasar
Centaurus A	13	22	32	−42	45	24	1330	Elliptical Galaxy–NGC 5128
3C 286	13	28	50	30	45	58	15	Quasar
3C 295	14	09	33	52	26	13	23	D Galaxy[b]
3C 348	16	48	41	05	04	36	45	D Galaxy[b]
3C 353	17	17	56	−00	55	53	57	D Galaxy[b]
3C 358	17	27	41	−21	27	11	15	Supernova remnant[a]–Kepler's supernova
3C 380	18	28	13	48	42	41	14	Quasar
NRAO 5670	18	28	51	−02	06	00	12	
NRAO 5690	18	32	41	−07	22	00	90	
NRAO 5720	18	35	33	−06	50	18	30	
3C 387	18	38	35	−05	11	00	51	

Name	Right Ascension hr	min	sec	Declination deg	min	sec	Intensity (flux units)	Identification
NRAO 5790	18	43	30	−02	46	39	19	
3C 390.2	18	44	25	−02	33	00	80	
3C 390.3	18	45	53	79	42	47	12	N Galaxy[c]
3C 391	18	46	49	−00	58	58	21	
NRAO 5840	18	50	52	01	08	18	15	
3C 392	18	53	38	01	15	10	171	Supernova remnant[a]
NRAO 5890	18	59	16	01	42	31	14	
3C 396	19	01	39	05	21	54	14	
3C 397	19	04	57	07	01	50	29	
NRAO 5980	19	07	55	08	59	09	47	
3C 398	19	08	43	08	59	49	33	
NRAO 6010	19	11	59	11	03	30	10	
NRAO 6020	19	13	19	10	57	00	35	
NRAO 6070	19	15	47	12	06	00	11	
3C 400	19	20	40	14	06	00	576	
NRAO 6107	19	32	20	−46	27	32	13	
3C 403.2	19	52	19	32	46	00	75	
3C 405	19	57	44	40	35	46	1405	D Galaxy[b]–Cygnus A
NRAO 6210	19	59	49	33	09	00	55	
3C 409	20	12	18	23	25	42	14	
3C 410	20	18	05	29	32	41	10	
NRAO 6365	20	37	14	42	09	07	20	Emission nebula[a]
NRAO 6435	21	04	25	−25	39	06	12	Elliptical Galaxy
NRAO 6500	21	11	06	52	13	00	46	
3C 433	21	21	31	24	51	18	12	D Galaxy[b]
3C 434.1	21	23	26	51	42	14	12	
NRAO 6620	21	27	41	50	35	00	37	
NRAO 6635	21	34	05	00	28	26	10	Quasar
3C 452	22	43	33	39	25	28	11	Elliptical Galaxy
3C 454.3	22	51	29	15	52	54	11	Quasar
3C 461	23	21	07	58	32	47	2477	Supernova remnant[a]–Cassiopeia A

The coordinates are given for 1950.

One flux unit = 10^{-26} watts/meter^2/hertz.

[a] Supernova remnants and emission nebulae lie within our own galaxy.

[b] D Galaxy refers to a Dumbell-shaped galaxy.

[c] N Galaxy refers to a galaxy with a bright nucleus.

Flux units—A measure of the amount of power being received from a radio source. 1 flux unit = 10^{-26} W/m^2/Hz. This means that if the area of the radio telescope (dish-shaped say) is 100 square meters and the receiver being used has a bandwidth of 1 MHz, then the power being received which is available to deflect the pen of the chart recorder is only 10^{-18} W if a 1 flux unit radio source is being studied. Present day radio telescopes, such as the NRAO interferometer can detect radio sources whose strength is only 10^{-3} flux units. Note that the deflection produced by the radio source as measured in units of antenna temperature depends on the size of the radio telescope. Flux units are the strengths of the radio sources themselves as measured by us on earth, while the antenna temperatures measured depend on the radio telescope being used.

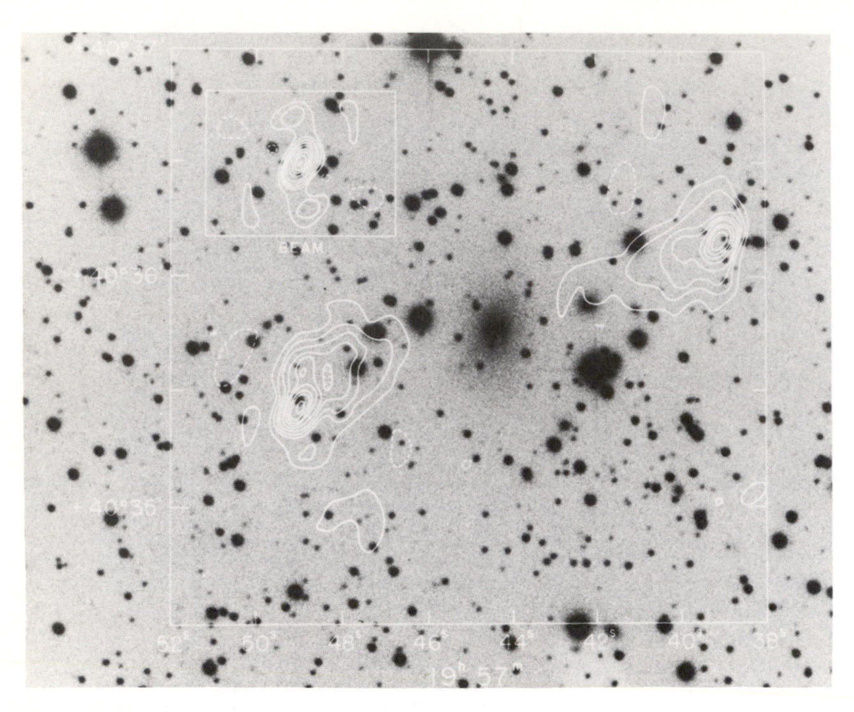

Fig. 43—A map of the radio galaxy known as Cygnus A mapped with the NRAO interferometer and superimposed on a negative of the star field. The central galaxy was once thought to be two colliding objects but is now thought to be a single galaxy undergoing violent explosive events. These explosions must have thrown out several blobs of matter, which are now emitting radio waves, and these are visible as the double radio sources seen in the map. Each part of the double, in turn, shows an additional structure indicative of possible multiple outbursts. (Courtesy NRAO.)

worked with the 200-inch telescope at Mount Palomar. His photograph of the region showed many galaxies and at the center, at the position of the radio source, a bright object quite unlike any galaxy he had ever seen. It looked almost like a double galaxy of some sort and Baade felt that the image might be produced by two colliding galaxies. There followed a debate over many years, concerning the likelihood of such an event occurring, and the tale is told how Baade bet Rudolph Minkowski a thousand dollars that the colliding galaxy hypothesis was the only way in which the object could be explained and the only way in which radio waves strong enough to travel through intergalactic space to our own Milky Way and still be detected, could be produced. At that time it was quite incomprehensible how an object so far away—and the Cygnus A radio source clearly was a very distant galaxy—could be a strong transmitter of radio waves (see Figure 43).

We are now so familiar with very distant objects being radio sources that we easily forget the stir this discovery caused. I still remember hearing late in the 1950s on the Overseas Service of the British Broadcasting Service the sounds of colliding galaxies, as they were then purported to be. The Service then played a recording of the noise produced by the radio telescope receiver and the increase registered when the telescope was pointed at Cygnus A.

In terms of the strength of the radio signals generated at the source, Cygnus A is far from being the strongest emitter in the universe, and we now know that whatever the cause of the radiation it certainly has nothing to do with collisions between galaxies. Rather there are even more incredible events taking place—explosions of whole galaxies, which are tearing them to pieces destroying all their stars in the process. We shall consider such explosions in more detail in the next chapter.

How Far Are Radio Sources—The Redshift

Well before the beginning of radio astronomy, the famous astronomer Edwin Hubble had discovered that the faintest galaxies, and therefore, presumably the most distant galaxies, showed the largest redshift. The redshift was therefore a measure of the distance to the galaxies. But what is the redshift?

We have discussed the emission of radiation, whether it be light or radio waves, at very definite wavelengths, by atoms or molecules. Optically, many atoms in galaxies emit their characteristic light signals at discrete wavelengths. The presence of these atoms is easily recognized by examining the spectrum of the galaxy. Hubble discovered that in order to positively identify the spectrum of giant galaxies he had to assume that the Doppler effect had shifted the lines a certain amount, and usually the interpretation necessitated the consideration of a motion away from us. In other words the spectrum was shifted to a longer wavelength, or the red part of the spectrum.

The redshift was larger and larger for fainter and fainter galaxies; hence redshifts came to be used as a distance indicator, since the fainter galaxies were also smaller in appearance and hence were probably farther away.

After astronomers identified some of the stronger radio sources such as Cygnus A and Virgo A, the redshifts indicated that the objects were unquestionably galaxies and very far away. Virgo A, an elliptical galaxy, is located about 33 million light years from Earth, and Cygnus A, is, according to its redshift, located at about 600 million light years away. This made the fact that we were nevertheless picking up strong radio signals from Cygnus A quite incredible. By comparison, our galaxy, viewed from that distance would never be detectable, using even the most sensitive radio telescopes.

CHAPTER 11

EXPLODING GALAXIES

Double Radio Sources

The first indication that Cygnus A was not necessarily a part of colliding galaxies came when better interferometric measurements showed that the radio object Cygnus A in fact consisted of a closely spaced pair of objects, neither of which was associated with the optical image in the center of the photograph made by Baade. Rather, each of the components of the double radio source lay on a line that passed through the optical object. By 1960, many radio sources were found to be double (see Figure 44) as the result of measurements with interferometers, particularly in England, which used increased baselines of the order of several tens of miles. Double radio sources were common, it seemed. But why?

Perhaps the light and radio waves traveled different paths through space! Perhaps the coincidence was fortuitous, but always a peculiar galaxy of some sort lay on the line joining the double.

Jets of Matter

In addition, some galaxies, notably Virgo A (Figure 45), were discovered to show long, very thin jets of luminous material protruding from them, and this had surely to be associated with explosions of some sort. But how could a large enough explosion occur in a galaxy that could eject a jet of material or two blobs of matter, detectable only by means of the radio waves they emitted, hundreds of thousands of light years into space?

We have mentioned that the center of our Milky Way emits strong radio signals, but if we placed the Milky Way at the distance of the Andromeda Nebula, i.e., about 1,800,000 light years, we would hardly be able to pick up its radio signals. Both Andromeda and the Milky Way are normal galaxies, in

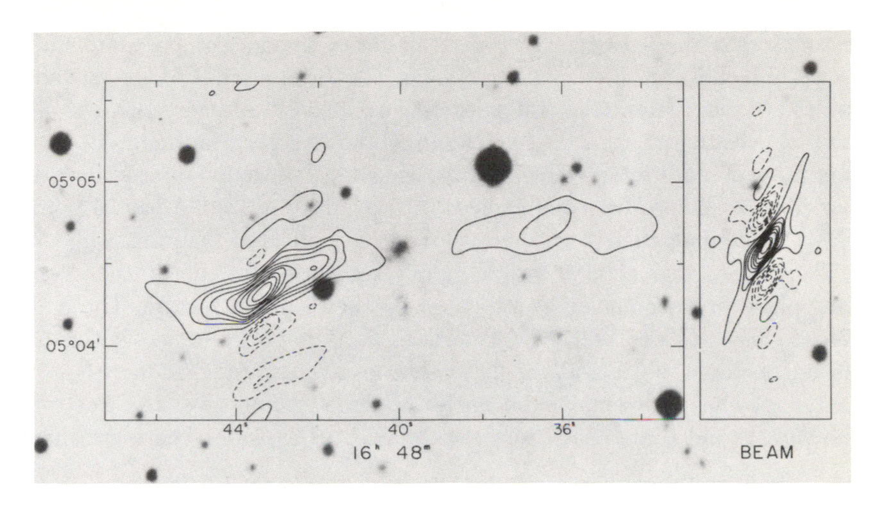

Fig. 44—The double radio source Hercules A in the Hercules cluster of galaxies mapped with the NRAO interferometer with the beam shape used indicated on the right. The radio map is superimposed on a negative of the starfield, and the radio galaxy itself can be seen between the two components of the double. (Courtesy NRAO.)

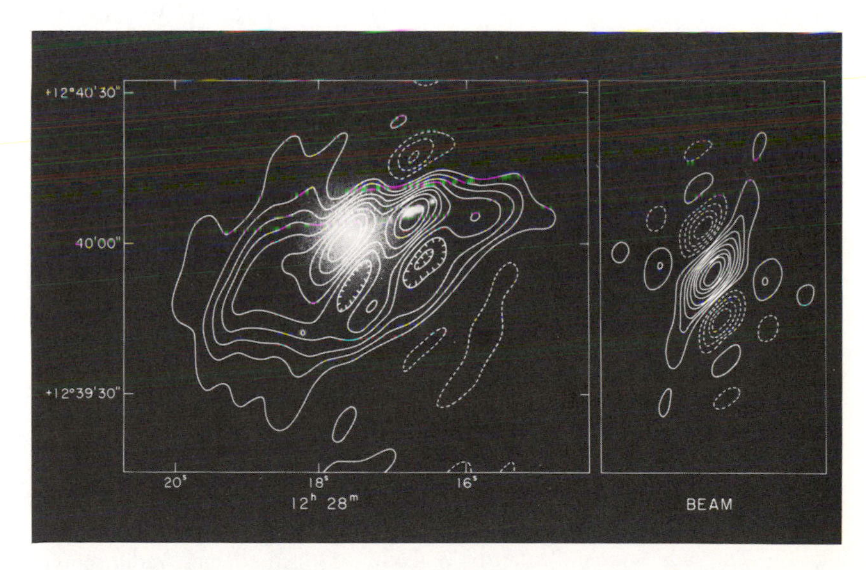

Fig. 45—A radio map of the radio galaxy M87 (Virgo A) made with the 3-antenna interferometer of the National Radio Astronomy Observatory. The small diagram on the right indicates the map that would have been produced if a radio source no larger than a point source had been observed. The fact that the brightest part of M87 shows radio contours similar to those on the right allows radio astronomers to state that there is a pointlike source of radio waves at the center of M87. The jet issuing from the galaxy also had radio emission sources associated with it. (Courtesy NRAO.)

the sense that they emit relatively weak radio signals compared to radio galaxies, which is the name given to objects like Virgo A and Cygnus A. Radio galaxies are most often elliptical galaxies, and often the largest member of a group or cluster of galaxies, which emit radio signals a hundred- or a thousand-fold more intense than normal galaxies. It usually appears that the radio emission comes from the nucleus of the galaxy or from a pair of points nearby. Sometimes their structures are more complicated than this.

There is another class of visible galaxy renowned for the drastic events taking place in their nuclei, as indicated by their optical spectra. These are called Seyfert galaxies, and violent motions are known to be taking place in their cores. Some, but not all, of these galaxies are also radio sources.

Perhaps chaotic events in the nuclei of some galaxies give rise to strong radio signals and sometimes cause the galaxies to explode. There are many

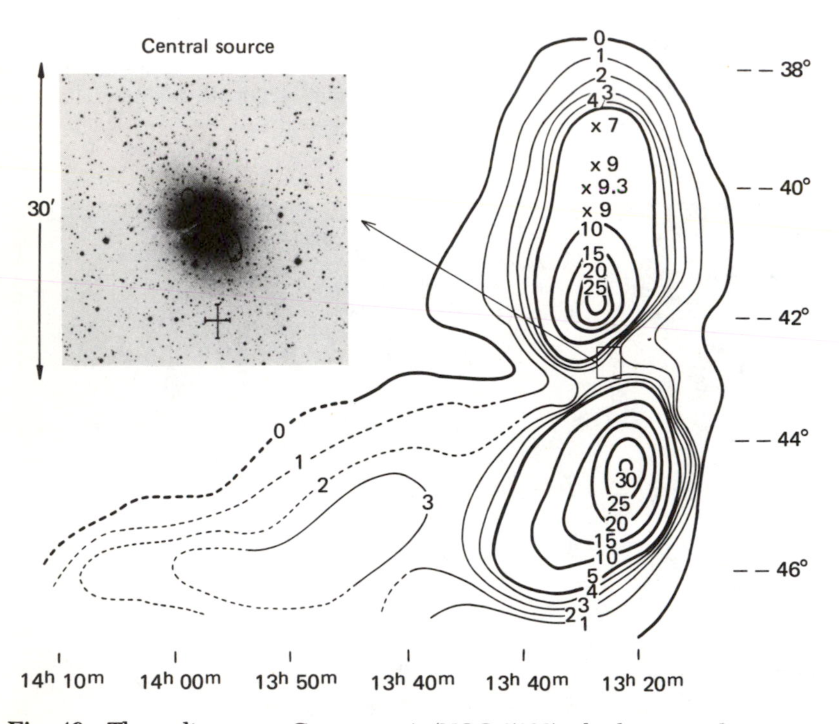

Fig. 46—The radio source Centaurus A (NGC 5128), the largest radio source in terms of its angular size. The small rectangle indicates the area for which the optical photograph is given, showing the galaxy, which must have undergone several explosions to have produced the enormous area of radio emission seen. The two small ellipses drawn in on the photograph indicate the location of another double radio source associated with this galaxy. The radio map was made with the 210-foot-diameter telescope at Parkes in Australia. The coordinates are right ascension and declination. (From *Australian J. Phys.* 18:589 [1965].)

double radio sources, usually associated with bright elliptical galaxies, which in turn usually show a very nonuniform appearance, not inconsistent with violent events taking place within them.

Explosion in Galaxies

The various types of objects mentioned above suggest that different degrees of explosive phenomena may be occurring in galaxies. In our galaxy there are motions that are relatively slow, only 50 km/sec away from the center as revealed by the hydrogen clouds, and the nucleus contains several weak radio sources. The double radio galaxy, like Cygnus A on the other hand, shows violent motions in its center, as do the Seyfert and more typical radio galaxies, the most dramatic example of which is Messier 82. M 82 can be seen to be exploding if its optical photographs are viewed in the right way. It is a strong radio source and there are tongues of matter sticking out, which are clearly moving at great velocities, according to the astronomers who have examined it very closely.

Another very nearby and extremely large-sized radio galaxy, Centarus A (Figure 46), a Southern Hemisphere object, shows evidence for several explosions having occurred at various times over the last tens of millions of years. Close to the optical object there is a bright pair of very small radio sources, but further out, as much as several degrees away, there is a series of very large radio sources clearly centered on the optical galaxy, which is famous for the dramatic dust cloud that cuts distinctly across its image.

Clearly then, explosive events must be occurring in galaxies. But why? And why do they usually eject two blobs of radio emitting matter in opposite directions?

Before we try to answer these questions, let us look at the quasars, which may be even more dramatic examples of cosmic explosions. If we could fully understand quasars, we might more fully understand radio galaxies.

CHAPTER 12

THE ENIGMA OF THE QUASARS

Many quasars are, according to their redshifts, even more distant than the most distant known galaxies (Figure 47). They also appear smaller in photographs than any galaxy, hence the name quasar—an abbreviation for quasistellar radio sources. The energy they emit is equivalent to all of the stars in the Milky Way, lumped together in a region a few light years across, having all their gravitational energy converted to radiation simultaneously—quite clearly impossible.

The Story of Quasar 3C273

The year is 1960 and there is one major problem that confronts radio astronomy; that is the nature of many of the radio sources which had not yet been associated with optical objects. During the previous 10 years many radio sources had been associated with well-known objects such as nearby galaxies or the remains of exploded stars in our galaxy or with hot clouds of gas between the stars. But there were many sources in whose direction optical photographs revealed nothing besides the usual conglomeration of normal-looking stars.

Perhaps some of the stars in the photographs were transmitting strong radio signals after all. But how could this possibly happen? We know that if the Sun, which gives out radio signals easily picked up on Earth, were placed at the distance of the next nearest star, we would simply not detect any signals from it. Clearly there had to be something emitting the radio signals which was not a normal star or was an invisible object, but how could radio astronomers pinpoint the radio source in order to decide between the two alternatives? We have already described how a radio astronomer can pinpoint the direction of radio sources by interferometry, and interferometers had been used during the 1950s to locate what turned out to be some of the more

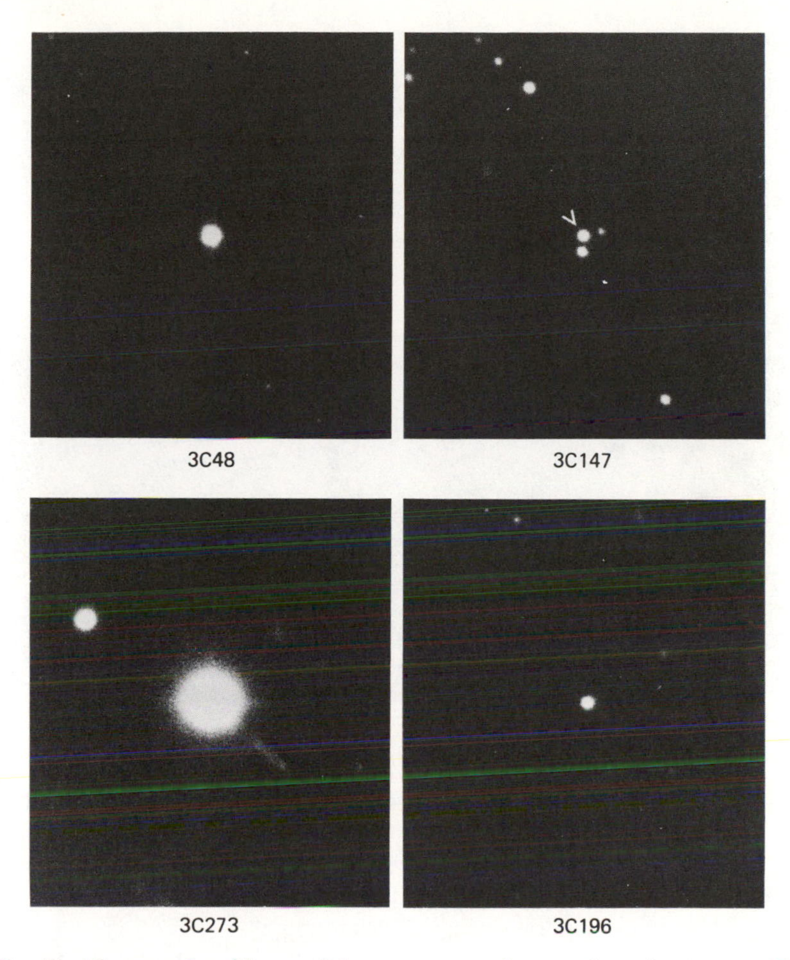

Fig. 47—Photographs of four well-known quasars that are also radio sources. 3C273 has a very obvious jet emitting from it, which looks similar to the jet associated with M87. (Courtesy Hale Observatories.)

obvious radio sources. They were therefore turned on the mysterious invisible sources.

Interferometer observations of a number of sources, in particular two called 3C48 and 3C273, indicated that both were in the direction, or nearly so, of starlike objects. The only other objects near them in photographic plates were also stars, so that even if the identification was uncertain, nothing much else could be done until the positional accuracy was increased many times.

So the concept of a radio star gained some currency in 1961 and 1962, although closer examination of 3C273 with bigger optical telescopes revealed a quite unexpected phenomenon associated with that "star." A jet of luminous material, 20 seconds of arc long projected from it (Figure 48). A star

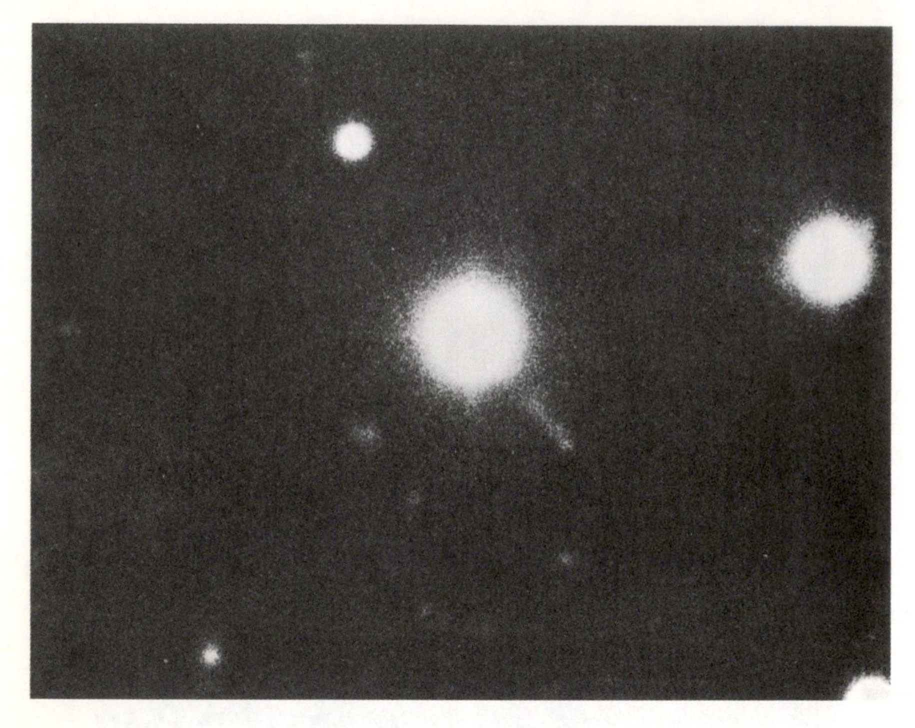

Fig. 48—Quasar 3C273.

with a jet? At least one radio galaxy, Virgo A, was known to have protruding jet of matter, but no star had ever shown anything like it. 3C273 could not be a star, but 3C273 was clearly not a galaxy either! It showed a small image in the photographs consistent with its being a star, whereas galaxies invariably showed a larger, more diffuse image in photographs.

What could be done about this? How could one become absolutely certain about the position of the radio source in order to make its identification more certain? There was one way which had never been tried before and that was to place a shield in front of the radio source, and if one knew exactly what the shape and position of the shield was, then one could derive the position of the radio source. But what shield could conveniently be located above a radio telescope on Earth?

A Radio "Star" Is Pinpointed

Finding a shield was simple: there already was one, the Moon! It moves slowly relative to the stars and will successively cover radio sources over a band across the sky. So in late 1962, a radio astronomer named Cyril Hazard pointed the giant 210-foot radio telescope at Parkes in Australia at the radio source 3C273 and waited as the Moon drifted slowly in front of the radio source. The time at which this was to occur had been roughly predicted and now the recording equipment in Hazard's lab showed this to be true. Suddenly the radio signals from 3C273 disappeared. The Moon had eclipsed the object, or as it is more correctly called, had occulted the radio source. Now Hazard had an accurate timing of when the radio source disappeared. But behind which part of the Moon had it gone?

This question could be answered, in part, by waiting until it reappeared. This 3C273 soon did, but the level of its signal first did not come back to the full value, but rather stayed at about half the original intensity for 10 seconds and then suddenly went up to the full value. The source that had reemerged from behind the Moon was a double object, consisting of two parts and yet the source that had disappeared had been a single object. How could this have occurred?

The answer, it turned out, was easy to visualize. Firstly, we know that the Moon has a circular shape. If the radio source were double and were lined up parallel to that small part of one edge of the Moon behind which it disappeared, both parts would disappear simultaneously, but on reemerging on the other side the curvature of the Moon there, relative to the double source, would be different and would allow first one part and then the other part of the radio source to be received again. And so it was for 3C273; it was a double radio source, the two components being separated by 20 seconds of arc. However, this was ascertained with certainty only after another occultation, because we should note that a single disappearance and reappearance of a radio source gives two possible positions in the sky. The Moon, of accurately known shape, moves in a well-determined way, and the time between disappearance and reappearance can be converted to a distance which the Moon traveled in the time and therefore the radio source has to lie on one of two lines drawn parallel to the direction of motion of the Moon. The timing of the disappearance and reappearance allowed the position on the two alternative lines to be calculated. The uncertainty is resolved by another occultation, perhaps weeks, months, or years later.

Hazard was lucky because another occultation of 3C273 was observable in Australia within months, and this enabled him to describe the position and structure of 3C273 with an accuracy comparable to that obtained in visual astronomy. It turned out that 3C273 was located at exactly the position of the starlike image and the second component lay precisely at the end of the jet!

Objects at the Edge of the Universe?

After the first pointlike radio sources to be accurately located in position in the sky were shown to be associated with optical objects, optical astronomers turned their undivided attention to them for the first time. Up to then, the optical astronomers had shown little interest in view of the fact that radio astronomers had never been able to tell them which optical objects were associated with the radio sources. The subsequent studies of the astronomers at such famous places as the Mount Palomar Observatory, with its 200-inch telescope, showed that these newly pinpointed radio sources appeared to be just like stars, even when viewed with the largest telescopes. That is, optical pictures showed them to be pinpoints of light with no extended envelopes of emission, which might indicate the existence of stars or spiral arms, which would have proved that they were galaxies. The exception, 3C273, showed the presence of the strange jet of matter. Perhaps it was some peculiar form of radio galaxy, whereas other apparently starlike radio sources, such as 3C48, 3C196, and 3C286, were stars.

How far away were these stars? To answer that question the astronomers had merely to observe the spectrum of the star and see which spectral lines were present, and once these were identified they could determine the Doppler shift and hence obtain a clue to the location of the star. If the Doppler shift was small it was simply a star in the galaxy, and the nature of the spectrum would help identify what type of star it was.

Maarten Schmidt at Mount Palomar was one of the first to study the spectra of these radio stars, but was soon stumped because they did not make any sense. The spectral lines were quite unlike anything he had ever seen in starlight. After pondering on this for some time he took the step required to solve the problem, and that was to consider if the spectral lines could not best be explained by the presence of elements commonly found in the spectra of galaxies, but showing a large Doppler shift.

This hunch proved true, and the spectral shift Schmidt found was a redshift, just as is found in distant galaxies, except that in this case the shift of the spectral lines to longer wavelengths was much larger than typically found for even the most distant galaxies. If one accepted that the universe was indeed expanding and that the redshift was a good indicator of distance, then the first of these pointlike radio stars to be studied had to be even further away than the most distant galaxies known at that time (1963).

Clearly these objects were anything but radio "stars." They could not be stars at all, but something quite different. Yet they were starlike in appearance! So the term quasistellar radio source was coined and abbreviated to *Quasar.* This did not solve anything of course; it just made talking about them easier!

How far away were these quasars? If they were at distances corresponding to their redshift, interpreted on the conventional picture of the expanding universe, then they were thousands of millions of light years away, but if they were that far away and we could still see them and pick up radio signals from them, then they were clearly enormously powerful radiators of energy—much more powerful than the most intensely emitting radio galaxy known at the time. What physical process could be occurring that would allow so much energy to be converted into radio or light waves? Possibly some explosive event such as could be seen in several radio galaxies, but these explosions had to be on a much larger scale than anything so far considered.

Maybe the quasar was not so far away. Then the redshift had to be produced by some other effect. Was this possible? The answer was yes, it was theoretically possible, but also very unlikely. It turns out that one of the predictions of Einstein's theory of relativity is that a wave trying to escape from a very dense object will suffer a redshift called a gravitational redshift. The wavelength is increased as the radiation tries to escape the enormous gravitational pull of such an ultramassive body. However, more careful thought on this showed that although the quasars could therefore be brought closer, and hence the energy problem reduced, the object would have to be so massive that the problem was as bad again! In any case another observation made the whole problem even worse for the theoretically inclined astronomer.

I should note that the radiation from quasars is almost certainly by the synchrotron mechanism discussed before, in view of the spectrum and polarization of quasars. The question is where do the electrons get their energy?

Variability on the Cosmic Scale

One of the most fundamental concepts in astronomy had been that, because most astronomical objects, apart from stars, were so big, they would be unlikely to show any variations over time scales that human beings might ever measure. For example, a galaxy might have a diameter of some 50,000 light years, which means that any disturbance, an explosion say, even if it traveled at the speed of light, would require several tens of thousands of years before it could illuminate the whole galaxy in question. It would take this long for the galaxy to change its overall brightness measurably, which means that we on Earth would never be able to observe any variation in our very finite lifetimes. The concept of variability of large objects in the universe was therefore never considered by astronomers.

One of the first quasars whose redshifts was measured was 3C273, and a search through old, carefully preserved photographs of the Harvard plate collection was made to study the appearance of 3C273 in years past. This revealed that the object appeared to have changed its brightness from year to year (over 80 years)—not always regularly—but the plates available showed that, by comparison with nearby stars in the picture, variations in intensity were definitely occurring. Any object that varies on a time scale of a year cannot be too much larger than a light year in size and this came as an added shock to those studying quasars. Not only did the quasars appear to be emitting more energy than is put out by all the stars in our galaxy, but the event, whatever it was, appeared to be confined to a volume of space of the order of a light year across. This conclusion holds, no matter how far away the quasars are, for even if they are close and we ignore the redshift, there is still an enormous amount of energy being generated in a very small volume of space.

One of the most intriguing problems that could now be tackled by astronomers was to find out just what the angular size of the quasars was. Knowing the angular size and a distance to the object makes it easy to calculate the linear extent (in light years or some other unit of distance). The variability observations had shown that one light year was an upper limit, because no data were yet available for determining if variations were occurring on a time scale of days or of hours. Since then variations over periods of several weeks had been found. There is a more direct way to measure the size of radio sources and that is to use interferometers. We discussed before the way a combination of two radio telescopes, separated by many miles, could, in principle, simulate a single dish of that size, so that it had the ability to resolve structure of the order of fractions of minutes of arc. It turns out that an interferometer can measure angular sizes of radio sources very easily, even though it might not be able to pinpoint their position too well.

If its redshift is interpreted according to the model of the expanding universe, 3C273 is about 1.5 billion light years from Earth. If it varies on a time scale of a year, then its diameter had to be a light year at most, and this would correspond to an angular size of 0.005 seconds of arc. The best optical telescope can measure angular sizes of only several tenths of seconds of arc. Could radio astronomers hope to better that? They could!

Long Baseline Interferometers

An interferometer in radio astronomy consists of two or more radio telescopes joined by cables such that the signals received by each of them can be combined at a central point in such a way as to produce an interference pattern. The use of the word "interference" in this context should not be confused with its use when referring to outside signals being received that interfere with routine astronomical observations. In the context of an interferometer we mean that the signals from the two telescopes are so combined that they sometimes reinforce one another and at other times cancel out one another, so as to produce an interference wave pattern as seen in Figure 42. This is the sort of pattern one expects if a radio source moves past that point in the sky toward which the telescopes are aimed, or if the telescopes are together driven so as to scan past a small, pointlike source of radio waves in the sky.

The farther apart the telescopes are (that is, the greater the baseline), the narrower each of the waves in the interference pattern is, when measured in terms of an angle on the sky. In fact this angle, the *lobe width* of the interferometer, depends on the separation between the two dishes and the wavelength being used, in just the same way that the resolving power of a single dish depended on its diameter and the wavelength being used. The farther apart the two dishes are, the smaller the detail that can be resolved, while the shorter the wavelength being used, the smaller the structure that can be studied.

Therefore, all that one needs to do to measure smaller and smaller angular sizes is to separate the two telescopes by larger and larger distances. However, one soon runs into problems after the telescopes are separated by more than several miles or so, because the cables needed to connect the telescopes become prohibitively long and they distort the interference patterns observed. This is because the radio signals running down these cables must not be delayed in any way; otherwise the interference pattern, or fringe pattern as it is usually called, is lost. This is because the cable lengths might alter slightly due to very small temperature changes, which delays the signal in reaching the receiver, and this produces the same effect as if the radio source being observed were moving slightly in the sky. Cables, even when buried underground for temperature stability, have limitations, and are also very expensive to install over such large distances.

The way to overcome this problem is to send the signal from a distant telescope to the base station by means of a radio link. In such a system the signal received by one telescope is converted to a signal at a convenient frequency for transmission by means of a microwave link of the type used for telephone communications networks. The signal is transmitted from the out-station to a receiving antenna near the main observing location and, in turn, control signals, for the movement of the distant telescope at the out-station, can be sent back across the microwave link. In this way two telescopes separated by hundreds of miles can be joined as an interferometer.

Now the limitation on the effectiveness of the system is the atmosphere, which effects the propagation of the signals through the link, because vapor, rain, or fog will affect the measurements, and when the distance becomes too great this method of connecting two telescopes as an interferometer also breaks down.

Another danger to such a system is the occurrence of unexpected obstructions in the path of the microwave link, which, of course, only works if one of the small dishes can be directly pointed at the other one, allowing line-of-sight transmission of the information. Jodrell Bank, in England, operated the first of the microwave link interferometers over distances of up to 100 miles. One day the signals from the remotely controlled out-station were lost altogether and nothing that could be done at the home site made any difference. There was nothing left to do but to go and check the microwave link repeaters on the way to the out-station and of course check the out-station itself. After several hours of driving across the plains and hills of England the astronomer found the trouble. A family, out for the day, had parked their camper directly in front of one of the microwave antennas and hence in the path of the signals passing through the microwave link! No sooner had they moved their camper a few yards forward than the interferometer was once again studying the sizes of quasars billions of light years away.

And so one can increase the baselines by taking the remote antenna farther and farther away, but what happens when one runs out of countryside in which to park the antenna, which is often constructed on some wheeled vehicle for easy transport? This happens fairly soon in England, so other methods had to be invented for doing the experiments.

Intercontinental Interferometers

The availability of very sophisticated tape-recording equipment, especially TV recorders, led to the use of transcontinental baselines and then intercontinental baselines. In such experiments each of the radio telescopes is equipped with a very accurate tape recorder and a superaccurate atomic clock whose pulses are needed to synchronize the recorders at the two telescopes. The idea is to observe a given radio source with each of the telescopes simultaneously and to record the signals on tape. The two tape recordings are then played back into a large computer at a later time, and the computer processes the data in such a way that interference fringes emerge, provided everything worked properly and the tape recorders remained running perfectly. Synchronization of the time pulses from the atomic clocks is achieved by reference to various worldwide navigation signals such as Loran C.

A large number of the very long baseline experiments has been made at the National Radio Astronomy Observatory in conjunction with groups from Caltech, M.I.T., and Cornell University. During the last several years baselines stretching from Australia to California, more than 80 percent of the Earth's diameter, have been used. Other baselines include West Virginia to

Fig. 49—A map indicating the sites used for some of the Very-Long-Baseline Interferometer observations. The length of the baselines is a measure of the equivalent diameter of the radio telescope that would have had to be built to make similar observations if the computer techniques and atomic clocks had not been available which allow these VLBI experiments to be done. (Courtesy N.R.A.O.)

the Crimea (in the U.S.S.R.), and Goldstone (in California) to the Crimea (see Figure 49). The latter system, operating at a 3.5-centimeter wavelengths, allows astronomers to measure radio source diameters as small as 0.0003 seconds of arc! This is an incredibly small angle, equivalent to the angle suptended by the letters on this page at a distance of 1500 miles!

The longest baselines have shown that there are structures in many quasars, and even in the nuclei of some radio galaxies, that are this small. This does not exactly simplify the problem of explaining the energy source in the quasars, or in radio galaxies for that matter, for the energies must be generated in very small volumes of space.

One of the problems in making observations using two telescopes, one in Russia and the other in the United States, has been to get the time accurately synchronized at the two places. The NRAO has built a completely portable unit for making tape recorder observations at any given radio telescope so that the equipment is standardized, but the time still needs to be checked. Exporting an unlikely looking crate containing very sophisticated electronic equipment to Russia, together with a very up-to-date, but considerably modified, TV tape recorder was a fascinating problem of logistics and topology for the radio astronomers concerned. In addition, getting back tapes rapidly from Russia to the United States to facilitate data reduction was a Herculean task. One need but try to imagine what an unknowing customs officer might make of a tape containing only radio noise from outer space, let alone a crate of electronic equipment used for studying radio signals from space!

The problem of getting the time to the remote telescope, especially in the Crimea, where the telescope stands at the edge of the Black Sea, is a fascinating story. Basically one has to synchronize a very sophisticated, but portable, atomic clock, with a Loran signal at some point at which the facilities are available, such as at the radio observatory in Sweden. The clock then has to be carried, with its battery power supply slowly using up its power, through several airports and customs officers, all the while plugging it into some convenient power source so that the batteries would not die. If they did, the intrepid astronomer would have to go back to the starting point. This clock had to have its own seat on the various flights and on arrival in Moscow would be met by one of the resident radio astronomers armed with a car for a journey to the other airport, whence on a plane to Simferopol. These journeys were always fraught with unlikely adventures, not to be repeated here. We have often speculated on the fate of a similar expedition of Russian astronomers carrying an enormous ticking clock in the plane beside them, occasionally plugging it into a wall plug at an airport, on a journey through the United States! It speaks very highly of the fantastic degree of cooperation that exists between two countries allegedly at odds with one another over so many things.

The experiments between Russia and the United States were successful and the reports describing the experiments appeared with the names of authors from both countries in the scientific journals of both countries. It appears that many quasars have angular diameters of 0.001 seconds of arc or less in diameter.

International cooperation provides us with a telescope effectively 7000 miles in diameter, and this had added to our knowledge of quasars, but the problem is still with us. Why do they emit so much energy?

Before I go any further I want to mention one fascinating sideline to the use of intercontinental baselines in interferometry. In order to use the interferometer properly one has to know the distance between the two telescopes very accurately. In fact when the Crimea-NRAO experiment was first performed a certain amount of difficulty was experienced, since the exact location of the Crimean telescope was not known. According to the coordinates for the telescope available to the radio astronomers at the NRAO and the maps of the Black Sea coast, the radio telescope was about a half a mile out to sea! However, with a little work and ingenuity, the computer was fed accurate enough data to allow it to analyze the data, and the map of the Black Sea area was corrected.

It turns out that if the position of some radio source is very well known, then the interferometer observations allow the baseline length to be calculated to an accuracy of a centimeter or less. Observations of two radio sources with two long baseline interferometers simultaneously, allow this experiment to be done with very high precision. There are some very interesting consequences of this type of measurement. One can actually measure such phenomena as continental drift or the change in the tilt of the United States as one end sinks slightly with respect to the other end. Other geophysical phenomena that will be measurable are the effect of tides on the shape of the solid Earth, the variations in the change rate of rotation of the Earth, and the wobble in the Earth's axis. These experiments are now being planned. But now, back to the quasars.

New Ideas

We still do not know whether quasars are very far away (at cosmological distances) or local objects, with their redshifts having to be interpreted as something besides a Doppler shift. If they are nearby, we have to explain why the galaxies nevertheless show a redshift, which appears to depend on distance, and why quasars do not appear to be associated with galaxies or clusters of galaxies in any obvious way. Halton Arp at the Hale Observatories (Mount Wilson and Mount Palomar) believes that he has evidence that all quasars are connected in some way to relatively nearby galaxies. For example, several quasars appear to be paired and centered about some peculiar galaxies and others appear to be connected by bands of luminous matter to other peculiar galaxies. If this connection is real—and not all astronomers believe it yet—then the large redshifts of the quasars have nothing to do with their distance and they must be something else. Several astronomers and astrophysicists are actively considering whether the redshift might indicate an entirely different phenomenon. For example, if one of the physical constants, such as the mass of the electron, or even Planck's constant, depended on time in some way, then the redshift would depend on the location of the quasar in space and on the time since its birth. If this were so, it would suggest the existence of a new law of physics of a form that we can hardly guess, at present. This seems the most attractive way out to me, but does not help solve the problem. However, an open mind is needed in this sort of work, and however reluctant astronomers might be to do so, we might have to consider this possibility more seriously, especially if the evidence grows that quasars are associated with galaxies.

One of the more reasonable explanations for quasars has been put forward by Sebastian von Hoerner, who has outlined a scheme in which the quasar is a very dense galaxy, so dense that there are a million stars crowded together in a region of space only one-thousandth of a light year across. That is a region only about a thousand times larger across than the Sun-Earth distance. If there were so many stars so close together, they would collide very often and these collisions would obviously destroy the stars in question, and they would produce flashes of light every time this happened. The theory predicts one collision every month on the average, which is consistent with the time scale of the variations seen in quasars. One might imagine that several collisions might occur near the surface and that these might throw matter out into space to produce the jets seen, although why the jets would be so long is difficult to understand. The energy problem is serious because when the stars collide in this way, they have to be converted to radiation very efficiently and this has not yet been proved to be possible.

In order to sum up and bring quasars into perspective I want to take you on an imaginary journey into space and see what we find.

CHAPTER 13

A JOURNEY TO A QUASAR

Let us take an imaginative journey into space. Let your imagination wander. Let it wander out beyond the Earth's atmosphere, outward past the Moon, past the planets, and past the Sun. Ever outward. We look back and can see the Sun, bright and alone, shining against the jeweled backdrop of the Milky Way.

Let your imagination travel faster and faster—at the speed of light. After a few hours we leave the Sun far behind and pass beyond the orbit of Pluto. Ever outward we go.

For days we travel; months and years go by.

At last another star, then another, and finally others loom in the distance. Some stars appear bright, others shine with a dull glow; some are red, some orange, some yellow. Some congregate in closely packed groups. As we pass one of these star clusters we can count hundreds of stars simultaneously wheeling through space. We travel onward. Clouds of incandescent gas appear but we soon leave them behind. Hot gas heated by newly born stars. As the years whisk by we pass hundreds of stars; many will have planets orbiting them, just like our parent sun and its solar system.

Onward we travel, slowly leaving the disk of our Milky Way Galaxy, so that it starts to unfold beneath us—a huge, spiraling, spinning, oscillating, overwhelming whirlpool of light consisting of hundreds of thousands of millions of stars. All drifting through space as part of this enormous galaxy.

The universe is now pitch black around us; only our home galaxy, still receding, is near us in space. We have left it behind, and it has taken our imagination several thousand years, traveling at the speed of light, to reach this incredible vantage point in space, which no man will ever reach except by a journey such as ours. If we had tried to travel from one end of our galaxy to the other, our journey would have taken 100,000 years! We leave behind us an incredible island universe, with its stars gravitationally locked together for billions of years.

Our galaxy, which seemed so huge to us, is now receding and appears ever smaller and smaller as we travel outward. We are entering the frightening void of intergalactic space. Another galaxy can be seen in the distance, together with several closely spaced companions in this lonely space. It is the Andromeda Nebula, which is soon also left behind.

As we travel even further out into the void we find that our parent galaxy is now visible as only a faint patch of light in the distance. But now we should speed up a little on our outward journey, for here there is nothing at all to see. Let us travel a hundred times faster.

Suddenly a whole cloud of galaxies looms out of the void ahead. It is a group of galaxies, containing tens of beautiful objects; some are spirals, like our own; some are compact, containing millions of stars clustered closely together with no semblance of the whirlpool pattern we left so far behind. Some small, some large, each with thousands of millions of stars, each star with its own planets; many planets have their own life forms. However, our imagination cannot comprehend that fully. There will be life forms similar to ours, and some will be unimaginably different, each living on its own little lump of rock, each a self-contained spaceship hurtling through space, endlessly, each unaware of the countless billions of similar rocks throughout the universe.

Let us continue to move onward, but faster this time. Maybe a million times faster. Then we see an unparalleled sight. A giant cluster of galaxies. Thousands of them tied together by unseen forces, thousands of spiraling, spinning, enormous groups of stars. To cross one end of this cluster to the other, passing hundreds of galaxies as we do so, would take several million years if we could travel at only the speed of light. We cannot start to count the number of stars and planets that might lie in this region of space. We cannot start to speculate how many lifeforms abound here. How many earthlike planets, how many like Jupiter, how many like Mars? Many are covered with teeming masses of life—creatures, insects, bacteria—all bent on survival in their own little world in this enormous universe. Each bent on survival under the rays of energy from their parent stars, some of which might explode suddenly, totally, in a brief holocaust, terminating all life processes on the planets around them.

And sometimes galaxies explode too. Some have already done so. Some of those we passed by in the distance were being torn asunder by forces that tear their structures apart, sending enormous clouds of matter out into space, generating signals so strong that man, back on Earth, can pick them up with his radio telescopes, his optical telescopes, and his x-ray telescopes, so that he can ponder the nature of the signals and try to figure out why and how such unbelievable events can take place.

Imagine, we are traveling through the unexplorable! We are now reaching parts of space beyond the reach of most telescopes through which fragile, fallible humans peer. We are reaching new regions of space. We still see the occasional cluster of galaxies hurtle by, but now space seems even emptier than usual. Then suddenly we notice a bright point of light in the distance. As we approach it, it becomes brighter. If it is a small galaxy, why is it so bright, and where are its companions in space? There are none! Where are the other cluster members? There are none! This object appears quite alone. It is very bright, but is very small. As we approach we see a faint jet protruding from it. As we move closer we realize that the jet extends an enormous distance into space. It is 100,000 light years long! And yet the parent object, if that is what the point of light is, still appears as only a point of light. There is no familiar spiral pattern suggesting that it might be a galaxy—no obvious cluster of stars to which we are accustomed.

The bright light comes from an object less than a light year in diameter! Perhaps it is a large star—a very large star? But no, stars are not that bright. Nor that hot, no matter how big they are. No, this point in space is radiating with the intensity of a hundred million stars and its signals have traveled one and a half thousand million years to reach Earth.

It looks like a star, but cannot be one. Quasistellar? Let us call it a quasar. A quasar is a strange object in space that fills a very small volume in space, but nevertheless succeeds in radiating as much power into space as if all the mass of the stars in our galaxy were simultaneously converted into radiation. But what is this jet out into space—protruding from an object so small? Impossible! But this universe we are traveling through is full of impossible things. Galaxies are impossible; so are stars, planets, life, humans, wars, love, hate, death.

And if we were to travel further we would find other quasars, some at the very edge of the universe perhaps? The edge of what? The edge of nothing!

We have now traveled 30 thousand million light years in our imagination. Here the quasars glow bright and alone in space, and most of them appear as single bright objects with no protruding jets. There are still galaxies around, but the quasars seem to avoid them and always outshine them. Look, there is another quasar, this one with a jet a million light years long. Again, quite impossible! That distance is the same as a typical distance between galaxies. But there it is. A starlike object, with a jet of luminous matter issuing a million light years into space. Thrown out from its nucleus by what unimaginable explosion on this cosmic scale. Who knows—perhaps the quasar will slowly rotate and start to wind the jet up like a spring and we would perhaps witness the birth of the spiral shape we see in so many galaxies. Perhaps the matter in the jet will cool and ultimately form stars. Perhaps other subsequent explosions will tear the galaxy to pieces again and start the processes all over. Perhaps matter is being formed here. Again our imagination has not been able to encompass all the possibilities, but perhaps, in who knows how many years time, we will find the answer to the riddle of the existence of the quasars.

In order to explain the quasars we appear to need some new laws of physics, but perhaps physics is blind or physicists are up a blind alley. Maybe astronomers are blind in overlooking some piece of data they already have? Is it all perhaps much simpler than we think? Maybe man was never destined to be confined to this solar system to study the universe out there. Perhaps journeys out there will be possible at some time in the unimagined furture.

Underneath our microscopes and beyond our telescopes lie phenomena that we do not understand at all. The quasars and the quarks may hold the keys to locks on knowledge, which we cannot yet comprehend. In the meantime, despite the enormous problems raised by recent discoveries in astronomy, the search for further truth about the contents of the universe and the nature of our existence goes on.

Probably all we need is one small probe and one small peek into a new direction and then . . . ?

CHAPTER 14

LITTLE GREEN MEN, WHITE DWARFS, OR NEUTRON STARS? THE STORY OF PULSARS

An Accidental Discovery

The discovery of pulsars is one of the most dramatic events to have occurred in astronomy research this century. An experiment, designed to study a phenomenon quite unrelated to pulsars was instrumental in detecting the incredibly regular signals that these strange objects emit. The signals occur in the form of a train of pulses, which had characteristics quite unlike anything that astronomers had expected at the time of their discovery in 1967. So strange, and yet regular, were these signals that the astronomers who were involved in the original accidental discovery were convinced for a time that they might have stumbled on signals from other civilizations in space. Such a discovery, if it ever occurs, is likely to be an accidental one, since many astronomers feel that we are more likely to pick up signals from some extraterrestrial communications network rather than signals specifically beamed at Earth.

But to return to the story of the discovery of pulsars. The scene was the radio astronomy station of the University of Cambridge in England where Anthony Hewish, with the assistance of a student named Jocelyn Bell, was engaged in an experiment to study the phenomenon of interplanetary scintillation at a frequency of about 100 MHz (a wavelength of 3 meters). This effect is similar to the phenomenon that occurs in the Earth's atmosphere and causes the twinkling of starlight, except that in this case Hewish was studying the twinkling of radio signals as they moved in a path through the solar system toward the Earth. The twinkling of starlight is produced by the effect of small irregularities in the Earth's atmosphere, which cause the light beam to bend and distort on its way to the eye such that the amount of light entering the eye varies irregularly. Similarly, the radio signals, in passing near to the Sun on the way to the radio telescope, pass through irregularities in interplanetary space. These irregularities are small clouds of particles emitted constantly by the Sun; some as the result of increased solar activity, some as a part of the normal solar wind phenomenon. Radio waves, in passing through these clouds, are bent and deflected, and on reaching the telescope exhibit a variation quite similar to the way the light waves vary in intensity on reaching the eye.

It is a property of this twinkling that it is only noticeable if the object being studied is pointlike so that only a single beam of light (or a single radio signal) is reaching the observer. If the object has a larger angular size, such as is the case with a planet as seen from Earth, then one finds that the many rays of light leaving the planet's surface travel through different parts of the atmosphere of the Earth on the way to one's eye. These rays average out so that the eye sees only a constant glow rather than the twinkling effect. For this reason planets, which are disk-shaped from our vantage point in space, shine with a steady light, but stars, which are pointlike from our vantage point, twinkle.

The same thing happens to radio waves from radio sources in the universe. If the radio source is a pointlike quasar, then it will twinkle (or scintillate, as this is called by radio astronomers), but if the object being observed is an extended object such as a galaxy or a supernova remnant, then the radiation being received will appear steady because all the scintillations due to many small parts of the source will average out to a steady glow. By studying the nature of the scintillations the radio astronomer can learn several things. Firstly, he can determine something about the characteristic size of the radio source he is studying; secondly he can learn something about the medium doing the scintillating, in this case the solar wind. The closer to the Sun the object is when it is being studied, the more it scintillates because the irregularities in the solar atmosphere are greatest near the Sun.

So at Cambridge astronomers set up equipment to study scintillations. This equipment was designed to detect very rapid changes in the signals being received, changes occurring on a time scale of fractions of a second. Most radio astronomical experiments cannot detect such rapid changes because it is customary to average the signal being received over some period of time, characteristically several seconds, in order to reduce the noise of the receiver and so facilitate the detection of weaker signals. The Cambridge group was therefore set up to detect very rapidly varying radio sources. That they did so with such startling consequences is now history, for what they discovered, in addition to many scintillating sources, were the rapidly pulsating sources now known as pulsars.

The first accidental observations of one of the pulsars was regarded as an unwanted source of interference on the records. Manmade interference, in the form of radar from a nearby airfield, or motor car ignition, has the characteristics of a string of pulses, either fairly regular, as in the case of radar, or irregular, as in the case of motor car ignition. So for many nights the experimenters noticed that they had a rather persistent source of interference at a certain time of night.

Miss Bell had, as part of her work, made a map of the sky showing the direction in which scintillating sources were seen. The telescope was not driven so as to be directed at any one point in the sky, but rather was fixed so that the sky drifted past the direction in which it was pointing. Jocelyn Bell noticed that one of the sources of interference, or one of the scintillating sources perhaps, was more or less overhead at midnight, a time when scintillation should be at a minimum, for then the radio waves certainly did not pass close to the Sun on their way to the telescope. However, the theory that an interfering signal had been picked up was also wrong because it was noticed that the time it occurred seemed to vary from day to day in a systematic way reminiscent of an object moving with the stars, rather than one fixed on Earth. Remember, Jansky's observations of radio waves from the Milky Way and the first accidental observations of Jupiter also showed this sort of time variation. It appeared that Bell and Hewish had perhaps discovered emission from some sort of flare star, stars that are known to flare optically at irregular intervals. However, in November 1967, one of these periods of radio emission was very strong and showed that there was a very definite string of regular pulses being received. Could these be radar signals from some other civilization? This theory could easily be checked by timing the arrival of the pulses on Earth, because if the source of the emission was on another planet, then the Doppler effect caused by the motion of that planet around its parent star would shift the apparent repetition frequency of the received signal in a well-determined way. The measurements of the period of the signals showed them to be constant to one part in 10 million, quite as good as most good clocks on Earth, but they showed no Doppler effect other than that expected for the Earth's motion about the Sun. By this time Jocelyn Bell had searched older data (3 miles of chart records!) and had found three more suspect sources of such pulsed radiation, and each of these was studied very carefully with another radio telescope which could track the objects and hence allow more detailed study.

All this work was carried out in great secrecy, especially at the point at which they were checking out whether the source of the strange radio signals could possibly be on another planet orbiting some distant star. The Cambridge group therefore kept their discovery secret for several months (much to the annoyance of many), and when they did announce the discovery, it set off an enormous flood of observations of these, subsequently named, *pulsars*.

In the early part of their study the Cambridge group named the sources LGM 1, LGM 2, and LGM 3, where LGM referred to "little green men"!

When it was discovered that the LGM's were not associated with stars, the Cambridge group tentatively suggested that the objects were in fact white dwarf stars, a late stage in stellar evolution; the death throes of stars in fact, during which time it was expected that the star might oscillate rather rapidly.

Subsequent study has shown that such a theory is untenable, and now the most likely object associated with the pulsed radiation is a neutron star, a star in which normal matter has completely degenerated so that it now consists entirely of neutrons, an object very difficult to imagine, but one which is now thought to exist, according to many theories about what happens to stars that die by explosive means.

Some Properties of Pulsars

The pulsars are so named because of the incredibly regular pulsed radio signals they emit (Figure 50). One of the first discovered emitted signals once every 1.33730110168 seconds! Any given pulsar, while always transmitting a constantly repeating signal, was not always detectable because the intensity of the pulses varied in a very irregular way, showing variations on nearly all time scales now studied (from seconds to years). This means that on any given day a radio telescope turned toward a given pulsar might not necessarily pick up any pulses. In addition it is now known that most pulsars are slowing down very gradually; that is, the repetition rate is changing very slowly, but also very uniformly. The repetition rates vary from one pulse every thirtieth of a second, to one every several seconds, for the 100 odd pulsars now known. The pulses can be received at all wavelengths, from a few meters, at which they were discovered, to a few centimeters. (See Table 3)

An important property of the pulses is that they are dispersed, which means that the radio signal from a given pulse does not arrive at exactly the same time at all wavelengths as observed on Earth. This is the result of the action of clouds of electrons that exist in space between the pulsar and the Earth. Even the clouds of neutral hydrogen gas, discussed before, contain a certain number of electrons as a result of the ionization of some of hydrogen atoms by cosmic rays or x rays, and in the path to pulsars, now known to be hundreds of light years away, there are enormous numbers of electrons. The effect of these clouds of charged particles is that the radio wave has a slightly different velocity at different wavelengths, due to the interaction between the wave and the particles. The longer the wavelength, the more the wave interacts with the particles and the slower it moves relative to the short wavelength radiation. This means that a pulsed signal, originally covering a broad range of wavelengths, all emitted simultaneously, will arrive on Earth at times depending on the wavelength. This phenomenon is known as

Table 3

A List of Pulsars Known as of June 1972, Showing Their Locations as Well as Several of Their Basic Properties

**********GO   FT06F001 ************PRINTER2****FMT <01>* *****144 LINES *** 1:28.07PM   7 AUG 73 CLARK   64 BERGER 64

YERVANT TERZIAN   CRSR AND NAIC   CORNELL UNIVERSITY   *PULSARS*   AUG. 7,1973   TOTAL NUMBER= 105   PAGE 1

PULSAR DESIGNATION	PULSAR	R.A. H.M.S.	DECL. DEG. ' "	L II DEG.	B II DEG.	PERIOD (SEC)	EPOCH (J.D.24+)	1/2 PULSE WIDTH (MSEC)	DISPERSION MEASURE (PARSEC/CM3)	PERIOD CHANGE (NSEC/DAY)	APP. AGE P/(DP/DT) (YRS)
P0031-07	MP0031	0 31 36	-07 38 26	110.4	-69.3	0.9429507566	40690	51.0	10.89	0.0366	7.1E+07
P0105+65	PSR0105+65	1 05 00	65 50	124.6	3.3	1.283651	41554	(29)	30		
P0139+59	PSR0139+59	1 38	59 45	129.1	-2.3	1.222947	41544	35	34.5		
P0153+61	PSR0153+61	1 53 00	61 55	130.5	0.2	2.351615	41554	(43)	60		
P0254-54	MP0254	2 54 24	-54	270.9	-54.9	0.4476	(41347)	10	10		
P0301+19	PSR0301+19	3 01 45	19 23 36	161.1	-33.2	1.38758356	41347	27	15.7		
P0329+54	CP0329	3 29 11	54 24 38	145.0	-1.2	0.71451864244	40622	6	26.776	0.177	1.1E+07
P0355+54	PSR0355+54	3 55 00	54 05	149.1	0.9	0.156383098	41701	3.5	57.0		
P0450-18	MP0450	4 50 22	-19 04 14	217.1	-34.1	0.548935369	41536	19	39.9		
P0525+21	NP0525	5 25 52	21 58 18	183.8	-6.9	3.7454934468	40957	181.0	50.8	3.461	2.9E+06
P0531+21	NP0531	5 31 31	21 58 55	184.6	-5.8	0.0331296454	41221	3	56.805	36.526	2.5E+03
P0540+23	PSR0540+23	5 40 10	23 28	184.4	-3.3	0.245965189	41701	10	72		
P0611+22	PSR0611+22	6 11 15	22 32	188.7	2.4	0.334918941	41701	6	96.7	4.84	1.9E+05
P0628-28	PSR0628-28	6 28 51	-28 34 08	237.0	-16.8	1.24441489598	40242	(50)	34.36	0.217	1.6E+07
P0736-40	MP0736	7 36 51	-40 35 19	254.2	-9.2	0.374918324	40221	22	100	<1.728	>5.9E+05
P0740-28	PSR0740-28	7 40 49	-28 15 16	243.8	-2.4	0.166750167	41027	8	80		
P0809+74	CP0809	8 09 03	74 39 10	140.0	31.6	1.2922412852	40689	41.3	5.84	0.0138	2.5E+08
P0818-13	MP0818	8 18 06	-13 41 23	235.9	12.6	1.23812810726	41006	22	40.9	0.1820	1.9E+07
P0823+26	AP0823+26	8 23 51	26 47 18	197.0	31.7	0.530659599042	40242	5.9	19.4	0.144	1.0E+07
P0833-45	PSR0833-45	8 33 39	-45 00 11	263.6	-2.8	0.0892187479	40307	2	63	10.823	2.3E+04
P0834+06	CP0834	8 34 26	06 20 47	219.7	26.3	1.27376349759	40626	23.7	12.90	0.587	5.9E+06
P0835-41	MP0835	8 35 34	-41 24 54	260.9	-0.3	0.76658	(41347)	20	120		
P0904+77	PSR0904+77	9 04	77 40	135.3	33.7	1.57905	(40222)	<30			
P0940-55	MP0940	9 40 40	-55 30	278.5	-2.1	0.664365	(41835)	30	145		
P0943+10	PP0943	9 43 20	10 05 32	225.4	43.2	1.097707	40519	50	15.35		
P0950+08	CP0950	9 50 31	08 09 43	228.9	43.7	0.25306504317	40622	8.8	2.965	0.0158	3.5E+07
P0959-54	MP0959	9 59 51	-54 37	280.1	0.3	1.436551	40553	50	90		
P1055-51	MP1055-51	10 55 47	-51 40	286.0	7.0	0.1971	(41347)		<30		
P1112+50	PSR1112+50	11 12 49	50 46	155.1	60.7	1.65643810	41594	18	9.2		
P1133+16	CP1133	11 33 25	16 07 33	241.9	69.2	1.18791116405	40622	27.7	4.834	0.323	1.0E+07
P1154-62	MP1154	11 54 45	-62 08 36	296.7	-0.2	0.40052	(41347)		270		
P1221-63	PSR1221-63	12 21 40	-63 47	300.0	-1.3	0.216475	(41835)		92		
P1237+25	AP1237+25	12 37 12	25 10 17	252.2	86.5	1.38244857195	40626	49.9	9.254	0.0325	4.6E+07
P1240-64	MP1240	12 40 20	-64 07 12	302.1	-1.5	0.38850	(41347)	60	220		
P1323-62	PSR1323-62	13 23 42	-62 03	307.1	0.3	0.529954	41733		313		
P1354-62	PSR1354-62	13 54 10	-62 13	310.5	-0.5	0.455780	41733		400		
P1359-50	MP1359	13 59 43	-50	314.5	11.0	0.690	(40585)	20	20		
P1426-66	MP1426	14 26 34	-66 09 56	312.3	-5.3	0.7874	(41347)	10	60		
P1449-65	MP1449	14 49 22	-65	315.8	-5.1	0.180	(40282)	(5)	90		
P1451-68	PSR1451-68	14 51 23	-68 32	313.7	-8.5	0.263376764	(40545)	25	8.6	<0.259	>2.8E+06
P1508+55	HP1508	15 08 04	55 42 36	91.3	52.3	0.73967787630	40626	10.3	19.60	0.433	4.7E+06
P1530-53	MP1530	15 30 23	-53 30	325.7	1.9	1.368892	40553	25	20		
P1541+09	AP1541+09	15 41 14	09 38 43	17.8	45.8	0.748448	(41347)	63	35.0		
P1556-44	MP1556	15 56 12	-44 31 30	334.5	6.4	0.25705	(41347)				
P1557-50	PSR1557-50	15 57 25	-50 36	330.7	1.5	0.192593	(41835)		270		

YERVANT TERZIAN   CRSR AND NAIC   CORNELL UNIVERSITY   *PULSARS*   AUG. 7,1973   TOTAL NUMBER= 105   PAGE 2

PULSAR DESIGNATION	PULSAR	R.A. H.M.S.	DECL. DEG. ' "	L II DEG.	B II DEG.	PERIOD (SEC)	EPOCH (J.D.24+)	1/2 PULSE WIDTH (MSEC)	DISPERSION MEASURE (PARSEC/CM3)	PERIOD CHANGE (NSEC/DAY)	APP. AGE P/(DP/DT) (YRS)
P1558-50	PSR1558-50	15 58 46	-50 49	330.7	1.3	0.864192	41733		165		
P1601-52	PSR1601-52	16 01 53	-52 36	329.7	-0.3	0.657959	41733		35		
P1604-00	MP1604	16 04 38	-00 24 41	10.7	35.5	0.42181507579	41005	15	10.72	0.0265	4.4E+07
P1641-45	PSR1641-45	16 41 07	-45 51	339.2	-0.2	0.454963	41733		449		
P1642-03	PSR1642-03	16 42 25	-03 12 30	14.1	26.1	0.38763877765	40622	3.7	35.71	0.154	6.9E+06
P1700-32	PSR1700-32	17 00 00	-32 40	351.7	5.4	1.211786	41852		103		
P1700-18	MP1700-18	17 00 55	-18	4.0	14.0	0.802	(41347)		<40		
P1706-16	MP1706	17 06 33	-16 37 21	5.8	13.7	0.65305345437	40622	12	24.99	0.550	3.3E+06
P1717-29	PSR1717-29	17 17 10	-29 32	356.5	4.3	0.620447	41554	(32)	45		
P1718-32	PSR1718-32	17 18 40	-32 05	354.5	2.5	0.477154	41554	(35)	120		
P1727-47	MP1727	17 27 53	-47 42 18	342.6	-7.5	0.829683	40553	30	121		
P1730-22	PSR1730-22	17 30 15	-22 29	4.0	6.3	0.871685	41853		45		
P1742-30	PSR1742-30	17 42 45	-30 40	358.5	-1.0	0.367421	(41835)		86		
P1747-46	MP1747	17 47 56	-46 56 12	345.0	-10.2	0.742347	40553	20	40		
P1749-28	PSR1749-28	17 49 49	-28 06 00	1.5	-1.0	0.562553168299	40127	5	50.89	0.705	2.2E+06
P1813-26	PSR1813-26	18 13 10	-26 50	5.2	-4.8	0.592887	41923		90		
P1818-04	MP1818	18 18 14	-04 29 03	25.5	4.7	0.59807262143	40622	10.3	84.48	0.545	3.0E+06
P1819-22	PSR1819-22	18 19 50	-22 53	9.4	-4.3	1.874393	41554	(76)	140		
P1822-09	PSR1822-09	18 22 40	-09 36	21.4	1.3	0.768543	41554	11	19.3		
P1826-17	PSR1826-17	18 26 15	-17 54	14.6	-3.4	0.307124	41554	(59)	207		
P1831-03	PSR1831-03	18 31 00	-03 40	27.7	2.3	0.686675	41554	(32)	235		
P1831-04	PSR1831-04	18 31 40	-04 31	27.0	1.3	0.2901071	41850		68		
P1845-01	JP1845-01	18 45	-01 27	31.3	0.2	0.659475	(41192)	(30)	90		
P1845-04	JP1845	18 45 10	-04 05 32	29.9	-1.0	0.59773452	40598	20	141.9		
P1846-06	PSR1846-06	18 46 07	-06 43	25.7	-2.4	1.451323	41554	(38)	152		
P1857-26	MP1857	18 57 44	-26 04 49	10.5	-13.5	0.612204	41026	25	35		
P1858+03	JP1858	18 58 40	03 27 02	37.2	-0.5	0.655444	40754	(170)	402		
P1900-06	PSR1900-06	19 00 30	-06 35	28.5	-5.5	0.431865	41554	(14)	180		
P1900+01	PSR1900+01	19 00 50	01 34	35.8	-1.9	0.7293016	41850		228		
P1906+00	PSR1906+00	19 06 45	00 35	35.1	-3.9	1.0169443	41849		111		
P1907+02	PSR1907+02	19 07 20	02 56	37.7	-2.7	0.494914	41554	(11)	190		
P1907+10	PSR1907+10	19 07 30	10 55	44.8	1.0	0.2836387	41851		144		
P1910+20	PSR1910+20	19 10 20	20 55	54.0	5.0	2.232964	41815		84		
P1911-04	MP1911	19 11 15	-04 45 59	31.3	-7.1	0.82593366503	(40624)	8.0	89.41	0.351	6.5E+06
P1915+13	OP1915+13	19 15 25	13 50 00	48.3	0.5	0.1946255	41274	15	94		
P1917+00	PSR1917+00	19 17 15	00 18	36.5	-5.1	1.272254	41554	(27)	85		
P1918+19	PSR1918+19	19 18 50	19 43	53.9	2.7	0.821034	41850		140		
P1919+21	CP1919	19 19 36	21 47 17	55.8	3.5	1.33730115212	40690	31.2	12.43	0.116	3.2E+07
P1920+21	PSR1920+21	19 20 36	21 08	55.3	3.0	1.077917	41850		220		
P1929+10	PSR1929+10	19 29 52	10 53 02	47.4	-3.9	0.22651703833	40625	5.3	3.176	0.100	6.2E+06
P1933+16	JP1933+16	19 33 32	16 09 58	52.4	-2.1	0.35873542051	40690	9.0	158.53	0.519	1.9E+06
P1944+17	MP1944	19 44 38	17 58 44	55.3	-3.5	0.4406179	(40618)	30	16.3		
P1946+35	JP1946	19 46 35	35 28 36	70.6	5.0	0.717306	40659	21	129.1		
P1953+29	JP1953	19 53 00	29 15 03	66.0	0.7	0.426676	40754	13	20		
P2002+30	JP2002	20 02 35	30	67.7	-0.7	2.111206	40756	15	233		

YERVANT TERZIAN   CRSR AND NAIC   CORNELL UNIVERSITY   *PULSARS*   AUG. 7,1973   TOTAL NUMBER= 105   PAGE 3

PULSAR DESIGNATION	PULSAR	R.A. H.M.S.	DECL. DEG. ' "	L II DEG.	B II DEG.	PERIOD (SEC)	EPOCH (J.D.24+)	1/2 PULSE WIDTH (MSEC)	DISPERSION MEASURE (PARSEC/CM3)	PERIOD CHANGE (NSEC/DAY)	APP. AGE P/(DP/DT) (YRS)
P2016+28	AP2016+28	20 16 00	28 30 31	68.1	-4.0	0.55795339053	40689	13.9	14.16	0.0129	1.2E+08
P2020+28	PSR2020+28	20 20 33	28 44 30	68.9	-4.7	0.343400790	41348	6.7	24.6		
P2021+51	JP2021	20 21 25	51 45 08	87.9	8.4	0.52919531221	40626	6.6	22.580	0.263	5.5E+06
P2045-16	PSR2045-16	20 45 47	-16 27 48	30.5	-33.1	1.96156082076	40695	79.0	11.51	0.945	5.7E+06
P2106+44	PSR2106+44	21 06 30	44 30	86.9	-2.0	0.414371	41845		129		
P2111+46	JP2111	21 11 38	46 31 42	89.0	-1.3	1.01468444509	41006	29	141.4	0.0620	4.5E+06
P2148+63	PSR2148+63	21 48 40	63 15	104.1	7.4	0.380141895	41701	15	125		
P2154+40	PSR2154+40	21 54 56	40 02 30	90.5	-11.4	1.52526334	41532	52	71.0		
P2217+47	PSR2217+47	22 17 46	47 39 48	98.4	-7.5	0.53846737644	40624	7.9	43.52	0.239	6.2E+06
P2223+65	PSR2223+65	22 23 30	65 22	108.7	7.0	0.682532	41833				
P2255+58	PSR2255+58	22 55 46	58 54	108.8	-0.7	0.368241	41554	13	148		
P2303+30	AP2303+30	23 03 34	30 43 49	97.7	-26.7	1.575884410	41006	26.0	49.9	0.2514	1.7E+07
P2305+55	PSR2305+55	23 05 00	55 26	108.6	-4.2	0.475063	41554	(27)	45		
P2319+60	JP2319	23 19 42	60 08	112.0	-0.6	2.25648387	41536	68	96		
P2324+60	PSR2324+60	23 24 00	60 55	112.9	0.0	0.2336518	41851		120		

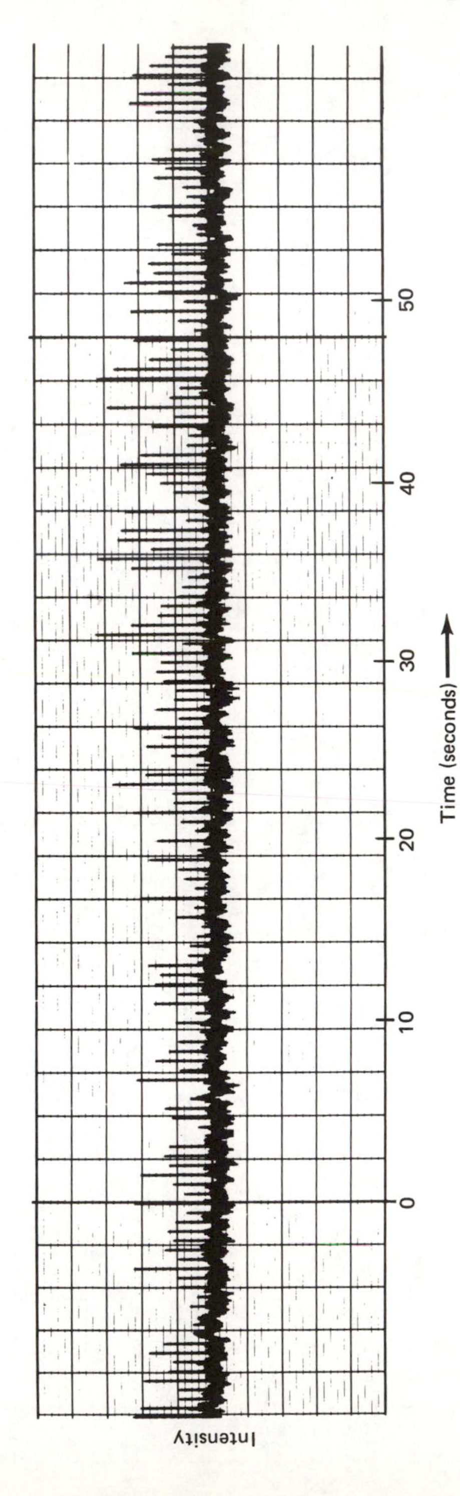

Fig. 50—A string of pulses from the pulsar PSR 2021 + 51 recorded on a paper chart. (Courtesy NRAO.)

dispersion, and it tells the astronomer much about the dispersive medium between his telescope and the pulsar. In fact, if he knows the distance to the pulsar, he can calculate the total number of electrons between it and the Earth by measuring the dispersion. That is simply the time delay in the arrival of the pulse at different wavelengths. Alternatively, if one knew the average number of electrons in space, one could determine the distance to the pulsar.

The Game of the Names

The original pulsars, when their discovery was made public, were referred to as CP 1919, CP 0950, CP 0834, and CP 1133. This designation of names for pulsars has reflected a desire by the radio astronomers to have their observatories remain known for their particular discoveries in the pulsar search. The second letter P, indicates a pulsar, and the first letter is usually the initial letter of the observatory at which the discovery was made. C indicates Cambridge, N indicates the initial letter of the NRAO, and JP would mean Jodrell pulsar. The following numbers indicate a position (the right ascensions) to astronomers. This practice of using many different letters has not survived, however, as 110 pulsars are now known and, to avoid confusion, most astronomers use a notation like PSR 1133+16, referring to pulsating radio sources located at right ascension 11^h33^m and declination $+16°$.

After the announcement of the discovery of pulsars (1968 to 1969) many observatories were involved in a frantic race to discover as many new pulsars as possible, but now the rate of discoveries has slowed down to a trickle as scientists study more carefully those already known in an endeavor to learn more about them. One of the most important clues in this search came in 1968 at the National Radio Astronomy Observatory.

Discovery of the Crab Pulsar

The observations of a pulsating star, or pulsar, in the direction of the remains of a stellar explosion have helped to shed light on both the nature of exploding stars and the pulsars.

As the result of a very sophisticated search technique using the 300-foot radio telescope at the NRAO, David Staelin and Edward Reifenstein succeeded in finding a pulsar at the position of the star that exploded in A.D. 1054 and now known as the Crab Nebula (see Figure 51).

Many other observatories all over the world had already been frantically engaged in searching the heavens for further pulsars, with some success, but

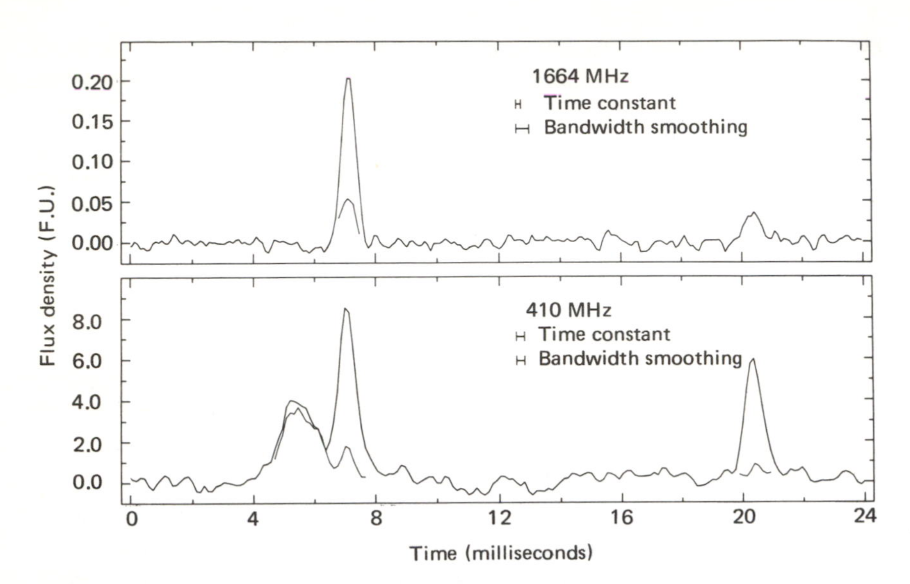

Fig. 51—Radio pulses from the Crab Nebula observed at two frequencies 1667 MHz (18-centimeter wavelength) and 410 MHz (75-centimeter wavelength). The pulses have been lined up by removing the effect of dispersion. The differences between the structure at the two wavelengths is very striking. The peak at time 20 at 410 MHz is called the interpulse. (Courtesy NRAO.)

the NRAO discovery has turned out to be the most important of all, since astronomers now know that at least some pulsars are the remains of the stars that exploded to form supernovae. This has been a disappointment to many who would have liked the explanation of pulsars to be more dramatic in terms, of say, an artificial object radiating as a beacon for spaceships or some other indicator of civilizations elsewhere. We will have to wait a little longer for such a discovery, however.

As with so many of the important advances in science, the discovery of the Crab pulsar contained a considerable element of luck. Staelin and Reifenstein, both on the staff of the NRAO at the time, were searching a strip of sky some 40° wide. They spent some 400 hours in their search program on the telescope. Their aim was to determine if any undiscovered pulsars existed in this strip of sky and their search technique involved the use of a radio receiver consisting of 50 separate data channels, all at slightly different wavelengths. This was the essential difference between their experiment and those done elsewhere. Others had used effectively only one radio receiver until then. Not only did the use of the 50-channel receiver of the NRAO enhance their chance of discovery of new pulsars, but it also allowed them to state with certainty whether they had found a pulsar or whether the effect had been caused by interference. This is a serious problem to radio astronomers, since interference consists of bursts of radio waves emitted by a whole host of manmade objects such as cars, airplanes, boosters on local TV antennas, or heater thermostats clicking on and off during the winter time. Sometimes distant FM or TV stations interfere with the radio astronomy projects at Green Bank, although the NRAO takes great care to prevent this from happening. The pulsars however, radiate most strongly in the wavebands that FM and TV stations use. For this reason radio astronomers tend to pick sites for their telescopes that are located far from manmade sources of interference.

The 300-foot telescope used 50 channels at slightly different wavelengths and it had already been established that the pulses of radio radiation from the pulsar arrived at Earth at slightly different times at different wavelengths. This is unlike manmade interference, which would appear in all 50 channels at the same time. This dispersion is well understood by scientists, and the equipment and subsequent computer programs for the reduction of the enormous quantities of data obtained during the search by Staelin and Reifenstein was designed to detect dispersed signals as well as periodic signals. Their data would not be as easily affected at interference as those of other observers elsewhere.

The Crab Nebula lay 2° inside the edge of the 40° strip of sky that was being searched. On October 17, 19, and 21, 1968, the two radio astronomers noticed that their data showed individual dispersed signals that had no fixed repetition rate, or period. It appeared to be neither like a pulsar nor interference. A pulsar would show both dispersion and period, and interference would show either only a fixed period or neither property (it would be random perhaps). It appeared that the astronomers were picking up isolated pulses from the direction of the Crab Nebula. They also found that two dispersion rates appeared to apply to their pulsar. This suggested strongly that they were picking up the radiation from two pulsars, possibly located close to one another in space.

They subsequently performed a separate experiment to measure the position of the pulsar, which allowed them to place one within the boundaries of the Crab Nebula and the other some 1½° in angle away from it. They also believed that since a number of the properties of the two pulsars, designated as NP 0532 and NP 0537, were so similar that they both were at about the same distance from Earth, suggesting that both were somehow associated with the Crab.

A Blinking Star

Another phase in the understanding of pulsars followed very soon after Staelin and Reifenstein announced their discovery. Many years before, in

1942, two famous astronomers, Walter Baade and Rudolf Minkowski, had speculated that one of a pair of closely spaced stars nearly at the center of the Crab Nebula was probably the remains of the star that exploded in 1054. Now that a pulsar had been discovered in the Crab Nebula it was obvious that an attempt should be made to determine whether this suspect star was indeed the pulsar, and this could be done by observing it with an optical telescope to see if the light from this star pulsated. It is ironic that the largest optical telescope in the world, the 200-inch at Mount Palomar, located for the favorable seeing conditions there, was prevented from observing the Crab Nebula until well after three or four other observatories in the United States had tried, because of the bad weather in California at the beginning of 1969.

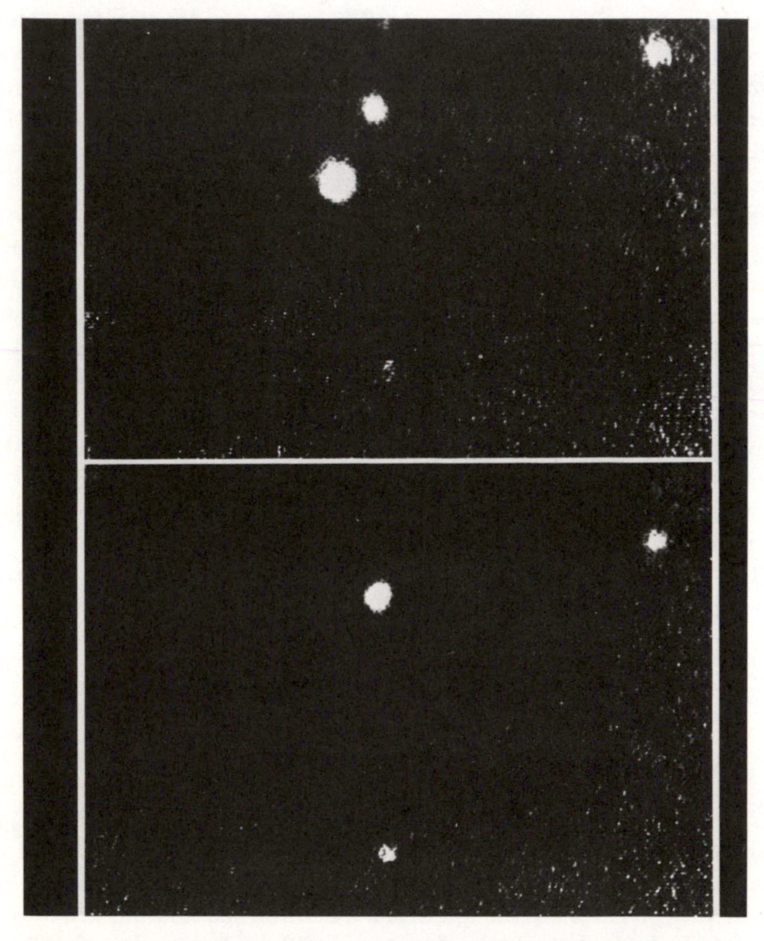

Fig. 52—Photographs of the pulsar in the Crab Nebula while on and off. (Courtesy W. J. Cocke and M. J. Disney, University of Arizona.)

At Kitt Peak in Arizona, two optical astronomers, Disney and Cocke succeeded in watching the star pulse. The fantastic thing is that the suspect star of Baade and Minkowski did indeed pulse at precisely the same rate as the radio pulsar; that is, once every three-hundredths of a second. Disney and Cocke succeeded in obtaining a photograph of the pulsar when it was on as well as showing that it did not radiate light signals when it was off, merely by opening and closing the shutter of their camera in synchronism with the pulsar (see Figure 52). Normal photographs are based on long-time exposures and therefore the pulsed light waves from this star would be averaged by the camera so that its pulsed nature had never been noticed before.

Here was final proof that pulsars were in some way associated with supernovae. The pulsar seemed to be a stage that the star entered after it had exploded and that part of the star that remained must have collapsed back in on itself. A search for optical pulsars at the location of the other radio pulsars drew a total blank, for no very obvious reasons, and indeed, a search for pulsars at the location of other supernovae also met with remarkably little success. Only two other pulsars of the more than 100 now known to exist appear to be associated with known supernovae. No others have yet been seen optically.

X-ray telescopes, flown in satellites and rockets, have detected x-ray emission from the Crab pulsar and the x rays, too, show the pulsed nature of the radiation. In fact several other pulsating x-ray sources have been found, but none of those appear to be associated with any of the known radio pulsars! Obviously much is to be learned about this phenomenon in years to come.

What Is a Pulsar?

The most popular opinion as to the nature of the pulsars is that they are neutron stars that are spinning very rapidly. Associated with these stars is supposed to be a very strong magnetic field, of the order of 10^{12} gauss. This magnetic field must contain some irregularity that allows radio waves to be generated at some unique point near the surface of these stars. The radio waves generated at this point must radiate outward in somewhat the way a searchlight beam does as its beam swings around and around. We therefore see a pulse only when the pulsar beam sweeps past the Earth. There may well be many pulsars whose beams are not now sweeping past the earth and which might be detected in the future.

The precise nature of the interior of a neutron star and the structure of its surface, which has been likened to a crystalline substance, is so difficult for the layman to envisage that one sometimes feels that it might be simpler to imagine that the pulsars are radio beacons belonging to other civilizations. Unfortunately there are now too many cataloged pulsars to make this a tenable theory.

There are also several additional properties of these bizarre objects with their pulsed signals, and that is that besides their regular pulsations, there appear to be changes in the precise structure of the pulses, which suggest other periodic phenomena occurring at the pulsar. If the pulse shapes are examined carefully it is found that there are drifting substructures and subsubstructures in the pulse shapes that can be examined by making very special observations. These have tended to confuse the picture rather than to clarify it, as is so often the case when more data are added to a little-understood problem!

Star Quakes

We mentioned earlier that the pulsars appear to be slowing down very gently. The rate of slowing down might well have something to do with how old the pulsar is. In addition, several pulsars, notably the one in the Crab Nebula, have shown sudden decreases in the period of the pulses, which then resume their gentle slowing down. Obviously something very violent must have happened to the object to make it change its rate of spin so suddenly.

It has been suggested that some matter has suddenly fallen into the pulsar, although the alternative—that the pulsar has changed its shape ever so slightly, as the result of what can only be described as a star quake—appeals more to the imagination. Such quakes have now been observed several times on the Crab pulsar.

Pulsars as Probes of the Interstellar Medium

We have already referred to the dispersion of the pulsar signals and how that effect is produced by the presence of charged particles in space between the pulsar and us. This allows the astronomer to study the distribution of the electrons in space, provided that he knows the distance to the pulsar. On the other hand, if he can determine the electron density, he can calculate the distance to the pulsar. In effect, the pulsar signals allow the astronomer to determine something about space, so that they are, in fact, a probe for studying the interstellar medium.

In addition, the pulsars allow the radio astronomer to study the interstellar magnetic field because the pulsar signals are polarized, and in passing along the path to the Earth these polarized signals are rotated by the

Faraday effect. The fascinating aspect of this is that the amount the polarized wave is rotated depends not only on the strength of the magnetic field but also on the number of electrons in the clouds in which the magnetic fields occur. Here then is an ideal opportunity for the radio astronomer to find the magnetic field strength because the dispersion he measured on the pulses gave him the total number of electrons to the pulsar (the distance need not be known, since he is simply measuring the total number) and the Faraday rotation measurement gives him the magnetic field strength times the total number of electrons. Provided that the field is associated with the clouds of electrons in some simple way, the astronomer need only divide the rotation measure by the dispersion measure to obtain the average magnetic field in space between the pulsar and the Earth.

Estimates of the field so found show that the interstellar magnetic field has a strength of about 3 micro-gauss. This is 10^{18} less in strength than the field at the pulsar! In other words, the pulsar itself has a field which is a million, million, million times stronger than the field between the pulsar and us, and yet we are able to measure the latter by studying the pulses in the right way and estimate the former, using some theory for the pulsar! It has been said that we have learned more about the nature of interstellar space as the result of the discovery of pulsars than we have about the pulsars themselves.

But what of the distances to the pulsars—how can we be sure just how far away they are? In the case of the Crab pulsar this distance is not in doubt, since there are many independent estimates for the distance to the nebula itself. The distance is 6000 light years. It is not yet known whether the companion pulsar is associated with it in some way. The dispersion of the pulses from the Crab pulsar leads to an estimate of the average density of electrons in the path to the Crab, which is 0.03 electrons per cubic centimeter, and the mean magnetic field between us and the Crab is 0.9 micro-gauss. Using this electron density for all space we can obtain the distances to the other pulsars.

There exists another way to determine the distance to pulsars that are not obviously associated with supernovae, and that is to study the effect of the clouds of neutral hydrogen gas on the radiation from the pulsars. Since neutral hydrogen clouds can emit radio signals around a 21-centimeter wavelength, they will also absorb at that wavelength. Therefore, if a radio receiver tuned to this wavelength is switched on and off, in synchronism with the pulsar, then one can study the absorption of the pulses by the gas. The reason that the receiver has to be switched on and off in this way is that while the pulsar is off the radio telescope would still be picking up emission from the hydrogen clouds and this signal would wash out the weak absorption being looked for. The point is that if absorption is found, then we know that the pulsar has to be behind the cloud (or clouds) of hydrogen, and an independent way exists for us to determine how far away the clouds are. Because the galaxy is rotating, we know that the Doppler effect on the hydrogen line signal depends on the distance of the cloud of gas being studied, and vice versa; if we know the Doppler shift and know the structure of the galaxy, we can calculate the distance to the cloud and hence to the pulsar.

Such measurements have been made on many pulsars, but they only allow the astronomer to say that the pulsar is beyond a certain distance. If he finds emission from an even more distant hydrogen cloud than the ones seen to be absorbing the pulsar signal itself, he can be fairly sure that the pulsar lies between these two clouds.

Clearly a lot of ingenuity goes into the study of the pulsed radiation from pulsars in order to determine their distances and learn most of what we now know about the electron densities and magnetic field strengths in interstellar space, not to mention into conjuring up coherent models of what a pulsar is.

Pulsars and Their Motion Through Space

In a recent experiment, radio astronomers were able to find out just how the pulsar is moving through interstellar space. The experiment was done by using two radio telescopes, one at Jodrell Bank in England, the other at Penticton in Canada. The pulses from pulsar PSR 0329+54 were simultaneously monitored by the two observatories for a 24-hour period, and the pulses studied with a computer which correlated the signals received. The experimenters found that PSR 0329+54 is moving at 360 km/sec, a very great speed for any object in interstellar space. This discovery probably helps to account for the fact that only 3 of the 100 or so pulsars now discovered appear to be associated in position with supernova remnants, although the pulsars are supposed to be the remains of the stars that collapsed after explosions. It appears that these stars can be thrown out of the supernova explosion at these very high speeds so that when we detect an old pulsar it has already moved a long way from its birth place.

CHAPTER 15

RADIO STARS

We have discussed how radio astronomers, using giant radio telescopes and sensitive receivers, can tune into radio waves from outer space. The first signals, found in the mid-1930s, appeared to be coming from the bright Milky Way band of stars, and we now know that these radio signals originate as a result of processes occurring in the space between the stars. In addition to these continuous radio signals there appeared to be several points in the sky from which much stronger radio waves were coming. In the early 1950s, the approximate location of many of these individual sources of radio waves were known and were at first loosely referred to as "radio stars."

It turned out that some of the so-called radio stars were located at the same position as some obvious optical objects like the Crab Nebula, the Orion Nebula, or some distant galaxy, usually a peculiar galaxy rather than a simple spiral one. They were obviously not stars, and indeed many other radio sources, as they were subsequently called, were not easy to associate with optical objects because radio astronomers were not able to pinpoint the positions of the sources accurately enough to allow comparison with photographs. Subsequent lunar occultation observations showed that several sources appeared to be located at the position of very blue stars. Again the term "radio stars" started to gain some ground, but these turned out to be what we now know as quasars.

We know of at least one star, the Sun, that gives out radio waves, but astronomers have showed that if the Sun were at the distance of the nearest stars it would not have been detected by our receiving equipment. However, certain stars are known to be weak radio sources; these stars are known to "flare," i.e., become very bright suddenly and then slowly return to their original brightness. Many hours of observations, with radio and optical telescopes observing simultaneously, were spent in the late 1950s and early 1960s monitoring these flare stars for possible radio emission. In a few cases weak radio signals were found which followed an optical flare on one of these stars, but the results were not very convincing to most astronomers. Besides these radio stars, no other ordinary stars were known to be, or were even expected to be, sources of radio signals that might be detected on Earth. However, using an interferometer, radio signals that could be positively associated with several nearby stars have recently been received, and these signals are carrying a strange message about our familiar neighbors in space such as Algol and Antares.

During 1970 the three-dish interferometer of the NRAO was used to track several stars with which all astronomers are familiar: Antares, Betelgeuse, Mira, Rigel and Aldebaran. The experiment was conducted in the off chance that the interferometer, which is the most sensitive and accurate instrument of its kind in the world today, might pick up something interesting. The radio astronomers, Robert Hjellming and Campbell Wade, also turned the telescopes onto a variety of other unlikely candidates; infrared stars, magnetic stars, hot stars, cool stars, shell stars, red stars, blue stars, x-ray stars, and novae.

The observations soon produced surprises. Antares was transmitting strong signals (strong compared to the Sun, but very weak compared to quasars) and two novae were also being received loud and clear. The latter two were Nova Delphini 1967 and Nova Serpentis 1970. It appeared that the

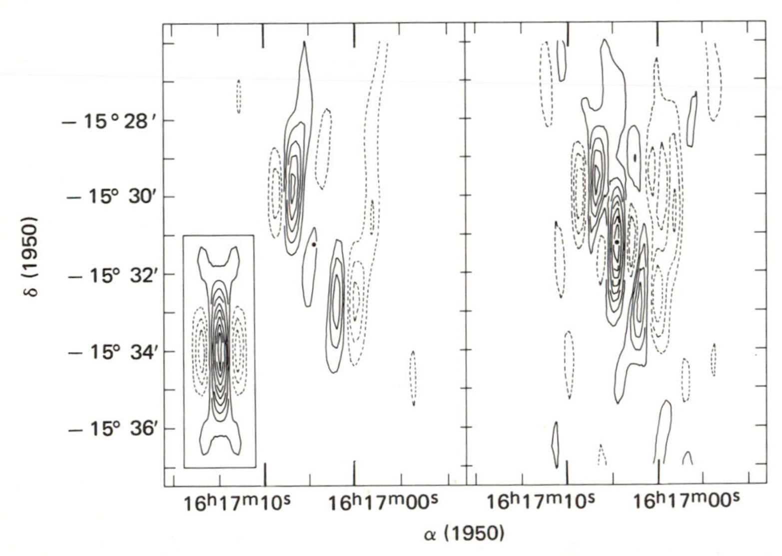

Fig. 53—A pair of radio maps of the radio source associated with the X-ray source called Sco X-1 on March 29 (a) and March 30, 1971 (b). For comparison the pattern that the antennas would have produced for a point source are shown in the inset. The dotted lines are so-called negative intensities. The black dot marks the position of the X-ray star, and it has obviously flared on March 30 while its companion double source has remained unchanged in strength. (Courtesy Westerbork Radio Observatory.)

radio emission was steadily increasing after the nova outbursts, probably from the shells of matter thrown off, which are still expanding into space.

The most interesting discovery that Hjellming and Wade made turned out to be radio signals from the first x-ray source, which had been discovered by x-ray astronomers. This source lies in Scorpius and is called Sco X-1. Earlier observations elsewhere had suggested that Sco X-1 was possibly a variable radio source. No one had seriously believed this until the Green Bank interferometer started regular monitoring of the object. Sco X-1 turned out to be wildly variable, sometimes changing its intensity by a factor of 60 in less than 24 hours, and sometimes being hardly visible at all. In addition, it appeared as a triple radio source, i.e., three points of emission—one at the position of the x-ray star (also visible optically) and the other two on either side of this (see Figure 53). Outside the solar system only pulsars (probably the remains of stars and therefore also radio stars in a sense) vary more dramatically than Sco X-1. At present it is difficult to prove that the two companion sources (2 arc minutes away) are actually associated with the x-ray star, since they emit a very steady radio signal and show no correlation with the properties of the x-ray star.

We do not really know why Sco X-1 is such a weird emitter of radio signals. Possibly cosmic rays are being thrown out into space in great explosions, or Sco X-1 may be a star rapidly rotating while winding up magnetic fields that stir up the surrounding gas regions so that radio waves are generated. While grappling with this problem, Hjellming and Wade turned up more startling discoveries. They knew that Antares was a binary system but at first were unable to associate the source of radio waves with one or the other of the two companions in space. Was the red supergiant, Antares A, the radio source, or was it the faint blue star, Antares B, 3.2 arc seconds away? Then one day (June 1, 1971), while Antares was being observed, a large radio flare was detected with the interferometer. This enabled the position of the source of radio waves to be measured much more accurately than had been possible until then, since the signal was much stronger and therefore the uncertainty in the measurement was much less. It was the blue star that had flared! However, the red supergiant is the one that is supposed to continuously spew matter into space!

During the same observing run, another x-ray source, this time in Cygnus, was found to be a radio source, and it also appeared to be directly associated with a binary star system. Clearly something very dramatic was happening in binary systems. The gas stream between the stars appeared capable of generating radio waves as well as x-rays, but did not always do so! Hjellming and Wade quickly got on the track of several other binary systems and they soon discovered that Algol and Beta Lyrae were also very variable radio sources, but that many others appeared radio quiet.

Since then, many of these objects have been constantly monitored, and recently Algol appears to have gone into a fit of periodic flaring, the flares occurring once or twice a day. The fact that no other binary systems have yet been discovered to be radio stars is very surprising, but perhaps the radio flaring does not continue year-in, year-out, and perhaps further observations will show if Algol will calm down again and if other binaries might, in turn, start up. No one even knows why such a phenomenon should be occurring.

One of the incredible things to emerge from the observations of Algol came from the examination of old and new optical data by Tom Bolton at the University of Toronto. He found that Algol was showing dramatic changes in its spectrum indicative of enormous changes in the amounts of matter that might be thrown out of that star, but up to now skeptical astronomers, used to studying good old dependable Algol, had simply rejected the older data as being worthless due to some spurious effects in their equipment. This event is becoming a set piece for describing how important it is to examine very carefully all the data obtained and not to reject something because it does not fit one's preconceived ideas.

No one yet knows why the blue dwarf companion of Antares A (500 light years away) is a radio emitter. No one knows why some binaries are radio emitters some of the time, why some x-ray stars are binaries and others apparently are not, why all binaries are not x-ray stars, or whether pulsars are related to any of these objects in some indirect way. Some pulsars also give out x rays, and indeed some x-ray sources are pulsed but do not give out radio or light signals! Much will be learned about these strange phenomena in the years to come and one of the most potent tools will be the new radio interferometer to be built by the NRAO in New Mexico. This interferometer, known as the VLA (for Very Large Array), is described in Chapter 17.

After 35 years of radio astronomy we are at last beginning to study the well-known stars in the radio band of the spectrum, and probably many more surprises are in store for us. Robert Hjellming predicts that the largest users of the VLA may well be optical astronomers wanting to know more about the common, "well-understood" stars that they have been studying all these years, but which now turn out to be anything but well understood.

Cygnus X3. What Is It?

On September 1, 1972, a radio astronomer in Canada, Phil Gregory, turned his radio telescope toward Cyg X3 (see Figure 54) because he had a half hour or so to spare before the radio star he wanted to observe, Algol, rose above the horizon. Up to then Cyg X3 had been a very weak radio source with an intensity of only one-tenth of a flux unit, in the radio astronomer's units of intensity. On this day, however, Cyg X3 had an intensity of 20 flux units—which at first appeared ridiculous, but after careful checking appeared

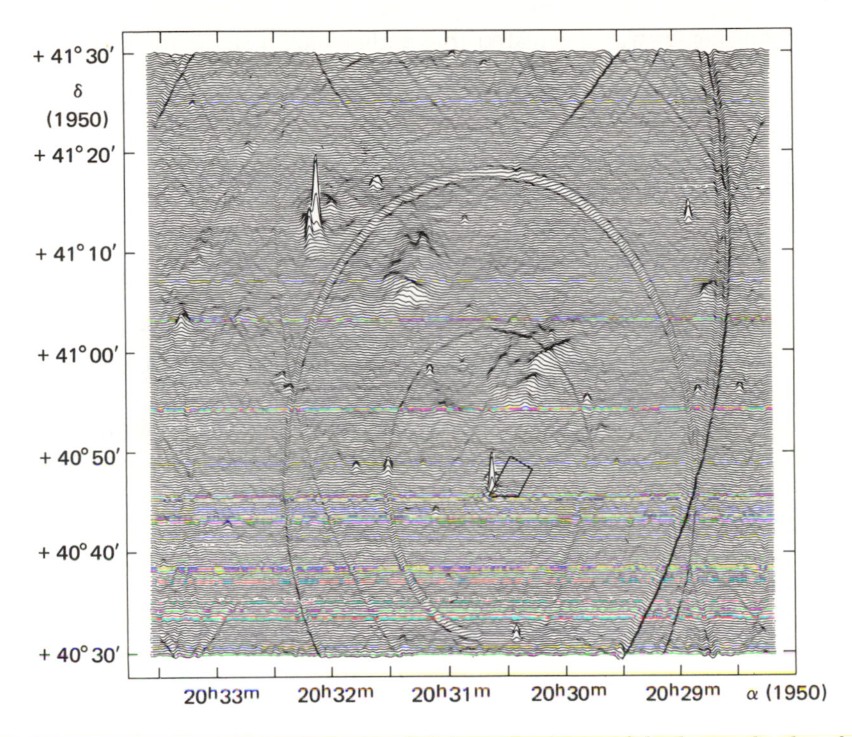

Fig. 54—A radio map made using a slightly unusual way of displaying the data for the region around the X-ray source Cyg X-3. The rings and parts of rings are known as diffraction patterns produced by strong radio sources in the map or beyond its boundaries and should be ignored. The quadrilateral marks the estimated position of the X-ray source as derived from satellite observations, and it is thought that the radio source visible on the edge of the box is associated with the X-ray source. (Courtesy Westerbork Radio Observatory.)

real. All the radio observatories over the United States, as well as several in Europe, were notified, and radio astronomers everywhere proceeded to monitor what appeared to be a giant explosion on this mysterious x-ray star, or whatever it is. Cygnus X3 is one of the many x-ray sources discovered by the Uhuru satellite and one of those subsequently detected as a radio source, albeit a very weak one, while it is totally invisible optically. Even during this radio flare it was still not visible, and even in the x rays it emits, it remained quiet. The measurements by the radio astronomers during this first event on Cyg X3 could best be described by the effects one expects to observe if an object explodes and throws out a shell of matter that expands into space and gradually emits weaker and weaker radio signals as the energy of the explosion leaks away. Before this object died completely, however, it underwent several more explosions over about 10 days. For these, however, the predictions of a simple model of a shell of matter being thrown out from the object did not seem to hold. This time a different physical process seemed to be occurring. It was speculated that rather than a single burst of energy driving out a shell of matter, it was possible that some slower, more continuous process was injecting particles into the space around this object.

Before one can seriously try to explain the nature of Cyg X3, it is necessary to know its distance. It turns out that such a measurement is possible, because when the object is bright in the radio waves it emits, it is possible to see absorption of these radio waves by intervening clouds of neutral hydrogen in interstellar space. From our knowledge of the structure of the Milky Way system, the distances to various hydrogen clouds are fairly well known, and therefore if one sees absorption of the radio waves from Cyg X3 by some clouds and not by others, then one can estimate the distance to the object. The French radio astronomers had the equipment on their telescope to make such a measurement during the first flare and derived a distance of some 11 kiloparsecs. Since then Cyg X3 has flared intermittently, and it is being very closely watched by radio astronomers the world over in the hope that they will learn the true nature of such strange objects.

CHAPTER 16

RADIO COSMOLOGY

Is the Universe Expanding?

Did the universe start with a big bang and expand outward or has it always appeared much as it does today? How big is it and how old is it? These are the questions posed by cosmology. It was hoped that radio astronomy would help provide the answer to these problems but at present the answers appear no closer. In fact, some radio astronomers now even question if the universe is expanding at all.

Cosmology is the study of the origins and scope of the universe, but this chapter will not be the usual discussion of how we learn about the expanding universe from radio astronomy observations. Instead, we shall examine how radio astronomy data affect our thinking in this field and how some of these data support a picture in which the universe may not be expanding at all. This may seem like heresy in the 20th century, but I feel that much evidence is gathering against a simple picture of an expanding universe. But before we do, let us consider how the concept originated and what the radio astronomer is able to contribute to it today.

In the 1930s, the famous astronomer Edwin Hubble showed that the faintest galaxies had the largest redshifts. We have already discussed the term redshift, which is the effect on light waves transmitted from an object moving away from us. The light appears redder to us than it was on leaving the object because the waves have been "stretched" in their passage toward us.

Since fainter galaxies are thought to be farther away, this discovery of Hubble's indicated that the more distant galaxies were moving away from us. In fact the speed of this recession seemed to increase as one examined fainter and fainter galaxies. The universe, therefore, appeared to be expanding with a speed that was greater in the past. This led to the concept of a "big bang," which started the explosion. To me this always has seemed a fundamentally religious picture of the universe. Many religions have in their dogma the concept of a moment of creation, and the big bang theory follows this pattern quite well. No one has ever explained what happened or existed before the big bang, which is essentially a unique event, almost Godlike!

If we are stuck with the fact that the universe is expanding, how do we avoid a big bang? The alternative is the steady state theory. In this theory the universe has always been and always will appear uniform, but is still expanding. Matter has to be created to fill in the holes as the galaxies expand outward. Before we discuss the validity of the expanding universe concept, let us see what radio astronomy can do for us. It turns out that, with even modest radio telescopes, radio astronomers should be able to pick up radio signals from galaxies far beyond the limits to which even the 200-inch telescope at Mount Palomar can see. If this is so, then obviously radio astronomy is a good tool for studying the universe on a much larger scale than optical astronomy can, and hence some fundamental cosmological questions, such as how big or how old the universe is, might be answered.

A simple experiment is to count the number of radio sources in the sky. One needs to count the number to successively weaker levels of intensity and then see if the numbers he finds are a reasonable fit to a picture of an expanding universe. The basic point is that the farther away a radio source is, the weaker the signals we pick up, and since the strength of the received signal decreases inversely as the distance squared ($1/R^2$) (as is the case for light as well) and we include more and more sources when we look to greater and greater distances (the volume of space included in our search region depending on the cube of the distance, R^3), then the number of sources we expect to see at a given level of signal depends on the strength of the sources to the 3/2 power.

The point of all this is that if a radio astronomer plots the number of sources he finds on a graph showing number versus the intensity he measures, then he expects a line with a slope of 1.5.

This is not found, and from the slope's deviation from 1.5, it is concluded that in the past there were more sources than predicted; that is, the sources were closer together, which fits an expanding universe, big bang theory, very nicely.

However, several astronomers have recently pointed out that more sources in the past is a statement that only has meaning relative to how many strong ones there are now, nearby. It can be argued equally well that if the number of weak sources we see is a true indication of a uniform universe, then there are instead too few strong ones at present visible in our skies! It turns out that a deficiency of only about 50 sources in our skies, at the higher radio intensities observed, can equally well explain the data. This is only 50 out of some 10,000 sources in the study.

This alternative might not be taken seriously if it were not for the fact that radio astronomers can find no radio property of objects such as quasars or radio galaxies that correlates with the redshift of the objects in a way that might support the fact that the redshift is a true indicator of distance. To place this statement in historical perspective, it should be remembered that for optically visible galaxies, redshifts were larger for fainter, i.e., more distant, galaxies. Of course there may be other less likely reasons that a galaxy is fainter; for example, it might always be younger or smaller.

However, one of the important properties of radio galaxies that might help shed light on the problem is the double nature of these galaxies. Let us picture that the double radio structure truly indicates that an explosion has occurred at some time in the past. Clearly, if we now see the two component parts at a large distance from one another, compared to say a closely spaced double, then we might, without any other data, conclude that the closer-appearing double is further away, assuming of course that the explosions occurred at roughly the same time. That is clearly a ridiculous assumption, but if one has a sample of a hundred double sources and if he interpreted their redshifts as distances so as to derive a linear size, then he might expect the sources with the largest redshift, being the youngest (that is, they are seen at a time shortly after a big bang), to have the smallest size. However, no correlation of size with redshift is found. In fact, no correlation of anything sensible with redshift is found in the radio properties. This makes us wonder if redshift is a measure of distance, since such obvious properties as intensity of the received signals might also be expected to vary with redshift. It does not do so either.

Lastly in this brief section, we note that several quasars, and even a radio galaxy, if their distances are interpreted according to their redshifts, appear to be expanding at speeds greater than that of light. This is a result of very long baseline interferometer observations of these objects. One way to avoid the problem of speeds greater than that of light occurring is to place the object much closer than its redshift indicates. We in fact have to ignore the redshift as a distance indicator. This means that the redshift may have nothing to do with an expansion of the universe. Alternative explanations have recently been put forward for this phenomenon, which we will not go into at this point, and further observations will hopefully clarify the picture somewhat.

Other Causes for a Redshift

One of the more obvious reasons that a spectral line should suffer an apparent redshift had already been suggested by Einstein as part of his theory of relativity. If a photon is trying to escape the gravitational attraction of a very massive object, then its wavelength will change as seen by an outside observer. The wavelength is also shifted to the red. Therefore quasars may simply be very massive and yet nearby objects, and still show a large redshift.

However, the mass of such objects becomes enormous for the redshifts observed. The problem of the existence of such massive quasars is almost as complicated as the energy problem if they are distant objects. If the redshift were truly gravitational on a large scale, then why do the fainter galaxies show larger redshifts? Surely fainter galaxies are not simply more massive objects! They are not, since astronomers are able to measure distances by other methods to some of the galaxies in question, which are already sufficiently far to show the redshift effect. What else can be occurring to produce the apparent contradictions?

Time Variations of Physical Constants

The wavelength of light emitted by an atomic transition is related to the energy difference between the levels of the orbits of the electron about the parent nucleus by a very famous equation which states that the wavelength is related to the energy by means of a constant term called Planck's constant, h ($E = h\nu$). The larger the energy difference, the larger the frequency emitted. It is always assumed that Planck's constant, together with other famous constants such as the speed of light and the gravitational constant, never change with time.

There was no reason for us to believe they might change with time of course, but now we are faced with so many difficulties in the theory of quasars and cosmology that we might need to reconsider this idea. Perhaps some, or all, of the constants are changing with time. We would then expect to see an apparently expanding universe because the most distant galaxies and quasars are being viewed at a great time in the past (the light has taken so much longer to get to us), and at that time the physical constants would have been different. We therefore see an apparent redshift which increases with distance, not because the galaxy is moving away faster, but because it existed at a greater time in the past. The universe is therefore not expanding at all. But why, and how, can the "constants" of physics change with time?

Perhaps we are here on the edge of the discovery of a new law of physics that determines how the other known fundamental laws depend on time. It is my feeling that such a law must obviously contain time as one of its basic elements. Man is conditioned by his birth-death cycle to believe that time flows in some way from a beginning to an end with many successive intervals in between. Hence a big bang theory. But what if the universe is not expanding at all? Then there may be no beginning or end. There may be no time of that sort. Will this new law of physics somehow release us from this time trap in which we find ourselves?

Perhaps we should consider a universe that is neither expanding nor contracting, has no beginning or end, but in which the so-called physical constants vary with time (or is it a variation with space, or distance). Not many astronomers or cosmologists believe such a picture, but there is an increasing number of doubters of the simplistic picture of an expanding universe and all it entails. Since this discussion reflects my prejudice, I have not discussed cosmology along the lines that might be followed in a more conventional astronomy book. Rather I will state that I feel much of cosmology today is belief and that this idea of a stable universe with physical laws changing in some way to be discovered is also a belief and not yet verified experimentally. It is undoubtedly true that we know very little about the cosmological questions that have been posed so far.

The Consequences of a Nonexpanding Universe

One of the consequences of a static universe would be that quasars are at an unknown distance, since their redshifts are meaningless indicators of distance. Rather they would reflect the properties required by our new law of physics. They may be close, but then clearly the new law will not involve a time factor. In other words, the light from quasars has not been traveling for any different length of time compared to distant galaxies with small redshifts. The new law might therefore involve time as measured with respect to the object itself, say since an explosion has occurred, or whatever other meaningful occasion we might be able to recognize. As far as galaxies are concerned, the fainter ones may still be farther, but the redshift is not indicative of a motion away from us.

We now can return to the question of how old, or big, the universe is. Previously we would have answered that the age could be found by reversing the expansion now seen and calculate how long ago it was to a single big bang. The answer is that it occurred about 10^{10} years ago. As to the size—well, anyone's guess would do. In our picture there is no expansion away from anything or into anything. It never started and it will not end. And if I might speculate even further, perhaps a new law involving time will give some science fiction ideas about time travel new popularity.

How does radio astronomy fit into cosmology in view of my skeptical remarks? It is at present producing more and more data that cast more and more doubt on the big bang and other evolutionary cosmologies, and it will probably continue to do this until someone is able to propose an entirely new approach to cosmology; for example, proposing a new physical law whose consequences can be tested by astronomers.

To sum up this section on how radio astronomy observations might relate to cosmology, I want to quote the words of Ken Kellermann of the NRAO, who stated that "while we radio astronomers may claim to be able to see further out in the universe than anyone, unfortunately we don't know where we are looking, and we don't know what we are looking at!"

Three Degrees Everywhere

In the study of cosmology we are basically concerned with what we might expect to see when we look further and further into space; in other words, further and further back in time. Will we see more or fewer radio sources as we dig down to lower signal levels? Also, if the universe did start with a big bang, would we not somehow expect to see the big bang itself if we looked far enough into space; that is, far enough back in time. Well, a little thought will show that there is a limit to how far back one can look, and that limit is reached when the distant parts of the universe are expanding at the speed of light. But what about the universe at some time after the initial explosion? Could we perhaps see some remnant of this event?

The way to answer that question is to try to measure if there is any radiation pervading the universe which cannot obviously be attributed to many faint and very distant radio sources. For such an experiment, one has to make very difficult observations—observations to try to measure the absolute intensity of the received radio signals. That means finding out how much the intensity of the received signal is above the level one would expect if there were no signals present at all. In terms used by radio astronomers, we would need to find out how much above absolute zero temperature the radio waves from space appear to be if a telescope is pointed at an empty part of the sky—empty of known radio sources, that is. Such an experiment is very difficult to perform because one has to have a radio telescope of perfectly known properties, which will allow the radio astronomer to calculate how much of the signal being received is being lost in passing through the antenna to the recorder he is using.

Such an experiment was done by two scientists, Arno Penzias and Bob Wilson at the Bell Telephone Labs (Figure 55), the same place at which Jansky had made his discovery. Penzias and Wilson used a very specialized horn type antenna for their experiment and discovered that after they had corrected their observations for all possible effects (including the effects in the antenna and the Earth's atmosphere), they still had a residual signal of 3 degrees unaccounted for. This experiment was soon verified by others elsewhere and at different wavelengths, and the results suggested that the universe was indeed 3° hot everywhere. Penzias and Wilson appeared to have discovered radiation from the remains of the primeaval fireball.

There appears to be only one way in which one can interpret this phenomenon and that is to invoke an expanding universe picture in which the universe was once in a very hot, very dense phase. So here we are

Fig. 55—The "ear" of one of the world's most sensitive radio receivers is this 50-foot long horn-shaped antenna at Bell Telephone Laboratories space communication station in Holmdel, New Jersey. The cosmologically important microwave background radiation was observed with this telescope. (Courtesy Bell Telephone Laboratories.)

contradicting some of the points made in the previous section, in which we argued that the concept of the expanding universe might be in doubt. The fact that everything, everywhere, is presently bathed in this 3-degree glow has not yet been accounted for on any other picture. However, if one returns to the picture of an expanding universe, one is left with the rather difficult philosophical concept in which a sudden singularity (an event that only occurs once) exists which gives rise to the start of the universe. It can also be shown, by theoretical argument, that the universe must ultimately stop expanding and collapse back on itself to end in another singularity. It is in order to avoid having to deal with singularities like this that some astrophysicists are suggesting that a new physics may be called for, which will provide the clue to our understanding this difficulty.

Testing Einstein's Theories

Radio astronomers, it turns out, can perform a very important experiment that will allow one of the predictions of Einstein's theory of relativity to be tested. Not many experiments exist that can be performed easily that might allow his theories to be tested and compared with alternative theories proposed by others. One of the predictions of relativity is that a light or radio signal will be bent a certain amount as it passes near a massive object, such as the Sun. The amount of this bending of the beam of radiation is predicted to be a certain amount by the theory of relativity and alternate theories predict a different amount of bending. The difference expected between theories is only 7 percent, and the amount of bending of the beam is very small, so the experiment is very difficult to perform.

The way radio astronomers can help here is to measure the position of a quasar, using an interferometer to see how the apparent position changes as the quasar passes near the Sun, preferably one disappearing behind the Sun to reappear at the other side at a later time. Dick Sramek at the NRAO has been trying this experiment for several years. Each year in October, the quasar 3C279 passes behind the Sun, and the quasar 3C273 passes fairly close to the Sun at the same time. Radio measurements of the position of the two quasars with respect to one another allow an estimate of the effect of the Sun on the signals from the quasars to be made. There is one severe difficulty in the experiment, however, and that is that the radio signal is expected to bend as it passes through the solar corona in any case, and this bending (refraction), due to the effect of the dense clouds of hot gas in the Sun's atmosphere, is much greater than that expected from the relativity effect. Fortunately a way exists to isolate the two effects. The bending due to refraction in the corona depends on wavelength, whereas the bending due to relativity does not.

Therefore by making observations at two wavelengths widely separated, such as 3 centimeters and 21 centimeters, the two effects can be separately estimated. The three-dish interferometer at the NRAO can operate at two wavelengths simultaneously, so Sramek has performed the experiment successfully on a number of occasions. However, the results are still frustrating because the uncertainty in the measurements is of the order of 10 percent—and thus a 7 percent effect cannot yet be proven. During the next few years, experiments using more refined equipment, and simultaneously, the interferometer at Caltech, in the Owens Valley, might finally allow the theories to be proved one way or the other.

CHAPTER 17

RADIO TELESCOPES OF THE WORLD

The World Distribution of Radio Telescopes

There are roughly 100 radio telescopes in the United States, many of which consist of fields of antennas rather than large dish-shaped telescopes. There are some 42 dish-shaped reflectors with diameters larger than 20 feet, and 15 of these belong to the U.S. government as part of the space program facilities or various military establishments. The NASA/JPL deep space stations also have five 85-foot-diameter dishes and one 210-foot dish in other countries.

For comparison, we note that there are a staggering 151 dish-shaped antennas greater than 20 feet in diameter in other countries excluding the U.S.S.R. Ninety-six of those form part of one system used for studying the Sun. Of course, the numbers I give here do not include the radio telescopes used as part of the radar defenses of the United States, since the radar network uses dish-shaped reflectors, which are simply radio telescopes and could also be used for radio astronomy if they ever became available.

Some of the largest fully steerable radio telescopes belong to NASA and are used for satellite communications—in particular, programs such as the Apollo flights to the Moon. As the Apollo program is phased out, more of the observing time on these beautifully instrumented dishes becomes available for radio astronomers. The 210-foot-diameter dish at Goldstone in California is equipped with one of the best low-noise receivers on any radio telescope in the world.

The largest single-dish radio telescope in the world is the 1000-foot-diameter instrument at Arecibo, in Puerto Rico. This dish, consisting of metal mesh strung out into a spherical surface in a conveniently shaped valley in the mountains of Puerto Rico, was originally used mostly for planetary radar experiments and studies of the ionosphere by radar and was funded by the

Fig. 56—The radome covering the 120-foot-diameter radio telescope at the Haystack Observatory, which is now a national facility funded by NSF and operated by the Northeastern Radio Observatory Corporation (NEROC). (Courtesy M. I. T., Lincoln Laboratory.)

Department of Defense. The Arecibo dish is now under the direction of the Space Sciences Lab of Cornell University and is funded as a National Center by the National Science Foundation. The telescope is having a new surface put on so that it will operate at much shorter wavelengths than its old 50-centimeter limit. It should be good to a wavelength of about 10 centimeters in the near future.

The National Science Foundation has also taken over the funding of one of the other radio telescopes previously built by the U.S. Air Force, the MIT 120-foot dish at the Haystack Observatory near Cambridge, Massachusetts (Figures 56 and 57). This telescope is enclosed in a protective radome and was also mainly used for radar experiments, such as bouncing radar signals off the planets. It was also used for studying the distribution of the notorious needles launched into space and released in a high orbit about the Earth. It is said that that was the first time a haystack was used to look for needles rather than someone looking for needles in a haystack!

To continue our discussion about the largest radio telescopes in the world, we find that the next largest dish, which is movable, unlike the Arecibo dish, which is fixed, but which can track sources by moving the antenna at the focus, is the recently completed dish at Effelsberg near Bonn in West Germany (Figures 58 and 59). This telescope has a diameter of about 330 feet and is part of the equipment of the new Max-Planck Institut fur Radio

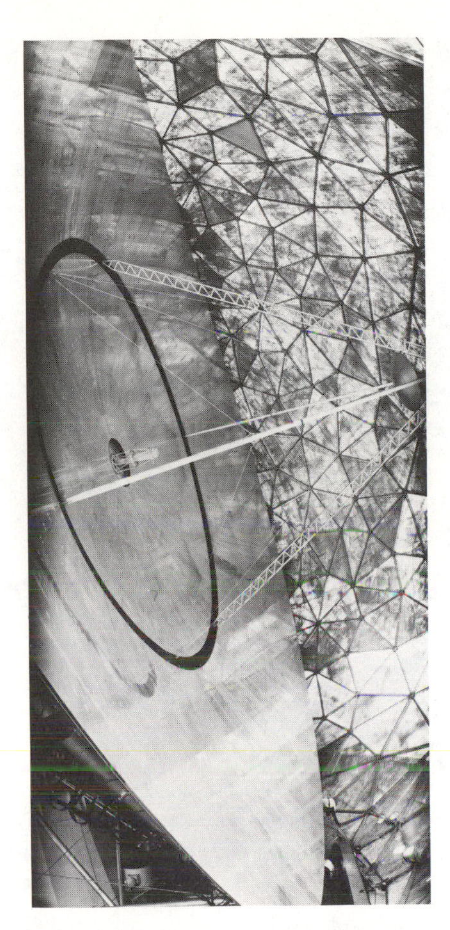

Fig. 57—The 120-foot-diameter antenna of M. I. T.'s Haystack Observatory, the most precise large moving structure ever constructed. It has a surface area of about a quarter of an acre. The accuracy of the antenna face is such that every point on the surface is within 0.1 inch of a perfect paraboloid. Notice the man standing near the bottom of the antenna. (Courtesy, M. I. T. Lincoln Laboratory.)

Astronomie. This dish was constructed with funds made available by the Volkswagen Foundation, and astronomers hold out high hopes for important new experiments to be done on it.

The next largest dish is the 300-foot transit telescope of the NRAO, in Green Bank, West Virginia, which has been in operation since 1962. It has been used very extensively at the 21-centimeter wavelength to map the neutral hydrogen in our galaxy, and the surface of this telescope was recently upgraded so that it can now work perfectly at the 6-centimeter wavelength.

The 250-foot radio telescope of the Nuffield Radio Astronomy Laboratories at Jodrell Bank in England is probably one of the most famous radio telescopes in the world, and has been in operation since 1957. Both Australia

Fig. 58—The 100-meter diameter telescope of the Max-Planck Institut für Radioastronomie, Elfelsberg, near Bonn. (Courtesy M.P.I.R.A.)

and Canada have 210-foot diameter dishes for radio astronomy work. The Australian dish is located at Parkes, N.S.W. (Figure 60). The Canadian dish is at Algonquin Park in Ontario and is operated by the National Research Council. Both telescopes are continually performing important experiments in radio astronomy.

Two of the somewhat smaller radio telescopes that have been performing extremely well over the last several years are the 140-foot-diameter dish of the NRAO at Green Bank, West Virginia, and the 36-foot dish of the NRAO located on Kitt Peak near Tucson, Arizona. The 140-foot telescope was completed despite initial difficulties in constructing the bearing on which the telescope rests, and it has completed much very important work since it was first used in 1965. Spectral lines from recombination of interstellar ionized hydrogen were first discovered with this dish, and the Zeeman effect at 21

Fig. 59—The opened receiver box of the 100-meter radio telescope near Bonn. The sensitive receivers are mounted just behind the focus in such a unit. The actual horn antenna is visible at the bottom of the box. Such receiver boxes, easily interchangeable, are used at many radio observatories. (Courtesy M.P.I.R.A.)

Fig. 60—The 210-foot dish at Parkes, Australia. This time exposure illustrates the full range of movements of the telescope as it is moved from its fully tilted position through 60 to zenith, rotated in aximuth by 180, then dipped in the opposite direction. (Courtesy C.S.I.R.O. Division of Radiophysics.)

centimeters was measured in 1968. Some 10 new molecules have been discovered with this telescope, and many of the other molecules discovered by radio astronomers were discovered using the Kitt Peak dish.

Some Large Interferometers

Many observatories in the world now have very sophisticated and sensitive interferometer systems. The one that was operational first on a routine basis was at Cambridge, in England, where much of the pioneering work on short-baseline interferometry was done. The word "short" here refers to distances of a mile or so. The Jodrell Bank group were pioneers in the use of intermediate-baseline interferometry (that is, baselines of 10 to 100 miles), while most of the very long baseline work is done in the United States and Canada. The Cambridge group at present has an eight-element (eight dishes) interferometer spread over several miles, which will make maps of radio sources and study the distribution of hydrogen in other galaxies.

Fig. 61—The Westerbork synthesis telescope. Twelve 25-meter dishes are arranged along a 1.6-kilometer-long line in an east-west direction. Ten are fixed in position and two are movable along a rail track 330 meters long. It is with this interferometer that the radiographs shown in this book were made. (Courtesy Westerbork Radio Observatory; copyright Aerophoto Eelde.)

The NRAO has its three-element interferometer (see Figure 1) operating on a routine basis, with the three dishes spread over a baseline of up to 8000 feet. The resolution of this instrument at a wavelength of 3 centimeters is 2 seconds of arc.

The Dutch radio astronomy observatory at Westerbork (Figure 61) recently brought into operation a beautiful 12-element interferometer that has produced probably the most dramatic results in the study of radio galaxies to date. The maps made of these distant radio sources with a resolution of 23 seconds of arc at 21-centimeter wavelength are all the more striking because the results are displayed in what look like photographs (called radiographs), and the maps can be more readily appreciated when studied by eye. These maps have shown that in several radio galaxies and relatively normal galaxies the radio emission, too, comes from spiral arms.

Aperture Synthesis

Radio interferometers that use more than two telescopes usually employ an observing technique known as aperture synthesis developed in Cambridge, England. By this method, a large telescope is completely synthesized by allowing the various smaller dishes to be successively located at every point over the area of the telescope to be synthesized. For example, let us picture two telescopes situated one mile apart. When the signals from these two dishes are combined, they act as two small pieces of a single dish with a mile diameter. If these two dishes were only 85 feet in diameter, the area covered by them would be very small compared with the area of a circle one mile in diameter. In fact most of the radiation from space falling on that large circle is lost. For this reason a simple interferometer has only a limited value; it can measure angular sizes, but cannot readily map the sources because virtually all the information has been lost.

There is a way around this and that is to use radio telescopes that fill in all the area of the one-mile circle. If one wanted to do that with 85-foot dishes, one would require nearly 4000 dishes. These would effectively simulate a single dish one mile in diameter. However, there is a way to simulate the effect of a large number of dishes and that is to move the dishes from point to point so that they sequentially are located at each of the 4000 locations within the large circle. Nature now steps in and makes the problem of doing this even simpler, because of the Earth's rotation. This acts in such a way that, as seen from space, one of the original two dishes sweeps out a circle around the other one in the course of a day. This means that by using two telescopes, one has already filled in the area equivalent to that of one telescope plus the area the other swept out in an annulus 85 feet wide around it.

If on the next day (or in fact 12 hours later, for the pattern of information one collects repeats after 12 hours) one of the telescopes is moved 85 feet closer to the first one, then one can wait another 12 hours as the rotation of the Earth sweeps out another annulus inside the one-mile circle. If one now stores the received radio signals in a computer for later processing, one can continue the process until all the area of the large circle has been covered by the combined effect of telescope moves and the Earth's rotation. One would require 31 moves of the 85-foot telescope to cover the large area, and only then can the data be processed to give results. The effect of a large dish one mile in diameter has, therefore, been synthesized. By using three or more dishes one can speed up the process of collecting the data, and fewer telescope moves are required to synthesize the larger dish. For example, with the NRAO telescope consisting of three dishes, and with the fact that information collected by any pair of telescopes is sometimes duplicated by another pair at a different separation, one finds that only nine moves of the telescope are required to make a full synthesis. This number of moves also depends on the size of the source itself, in that the smaller it is the fewer moves are required. We cannot go into the technical reasons for this in this book. The Dutch radio interferometer, operating on the synthesis principle, requires five moves of two of its 12 dishes, the others staying fixed in position, in order to make a complete map of a radio source (see Figures 62, 63, 64, 65 and 66).

The NRAO has recently been given the go ahead to start work on a large array of radio telescopes operating on the synthesis principle. This array, called the VLA for "Very Large Array," will consist of twenty-seven 85-foot dishes strung out along the three arms of a wye (Y), with each arm some 13 miles long, with the synthesized telescope having a diameter of 25 miles. No telescope moves are required and a complete map becomes available after eight hours of observations. This VLA is to be located in New Mexico, where the level of man-made interference is still very low and where the water content of the atmosphere is also very low, which will allow observations at very short wavelengths to be made (see Figure 67). We cannot even speculate about the dramatic new discoveries such a powerful instrument might bring to the astronomy world, not the least of which might be the usefulness for studying newly discovered phenomena such as radio stars and maser sources of molecular line emission, which are known to be very small objects. The estimated time of completion for this array is 1980 or 1981, and the resolution at the 21-centimeter wavelength will be one second of arc and at the 1-centimeter wavelength, $\frac{1}{20}$ of a second of arc. Figure 68 compares the type of maps that would be produced by various telescopes mapping the same galaxy.

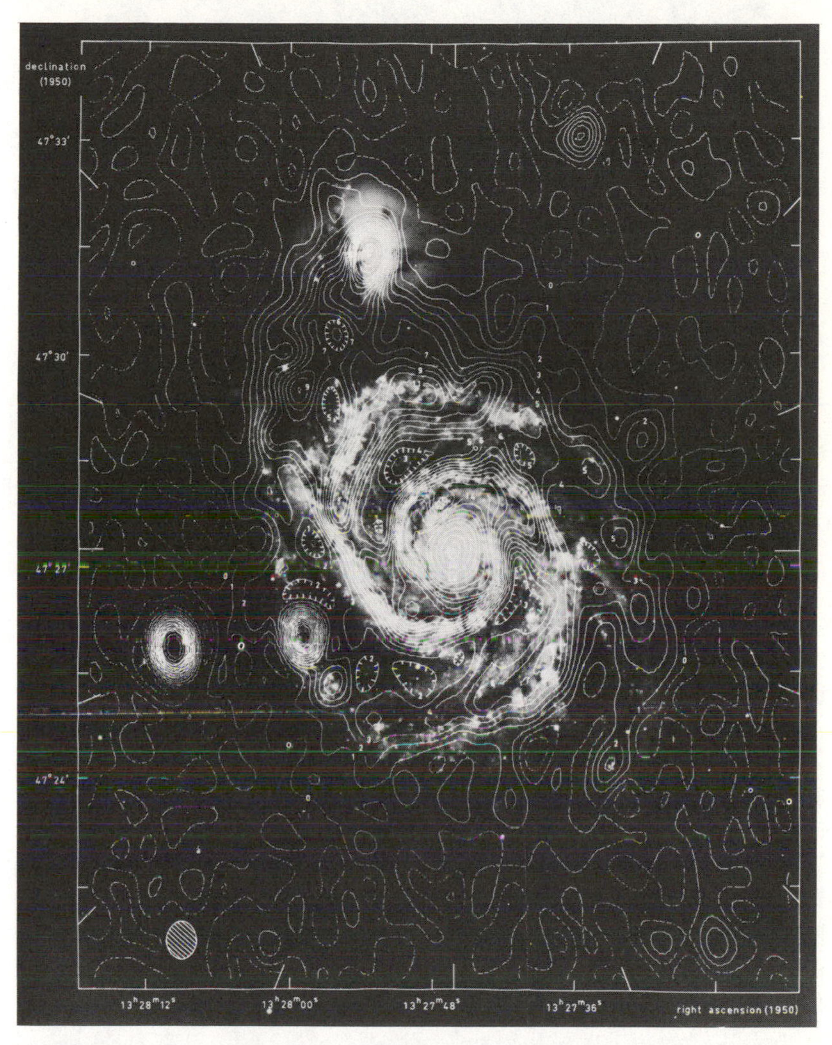

Fig. 62—A 21-centimeter wavelength map of the galaxy M51 made with the Westerbork synthesis telescope. The relatively strong source toward the left-hand edge of the map is probably a background radio source not associated with M51, while the other strong source, just to its right and lying on the optical spiral arm is probably a supernova remnant in that galaxy. (Courtesy Westerbork Radio Observatory.)

Radio Telescopes of the Future

Besides the VLA, at least one other large radio telescope is in the planning stage and that is the 375-foot telescope being planned by the University of

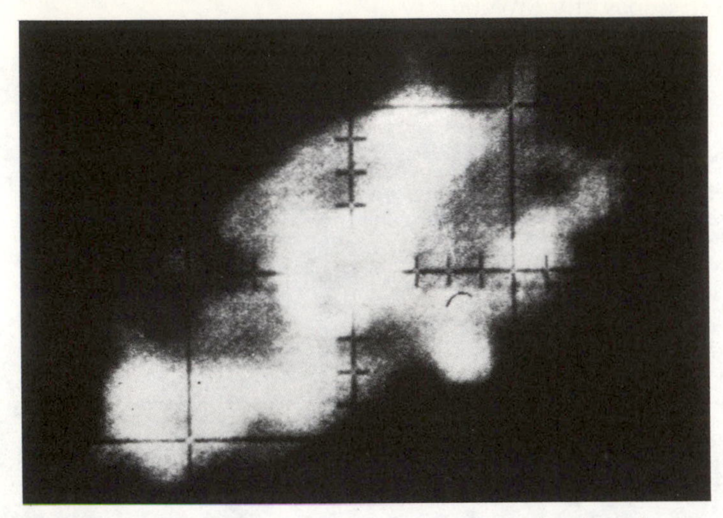

Fig. 63—The radiograph of a galaxy known as Maffei 2. This galaxy is virtually invisible optically, since it is obscured by dust clouds in the Milky Way lying in front of it. However, infrared radiation from this galaxy is clearly detected, and this resulted in its discovery by the Italian astronomer Maffei. The radio waves from this galaxy can also penetrate through the dust, enabling this radio photograph to be made. (Courtesy Westerbork Radio Observatory.)

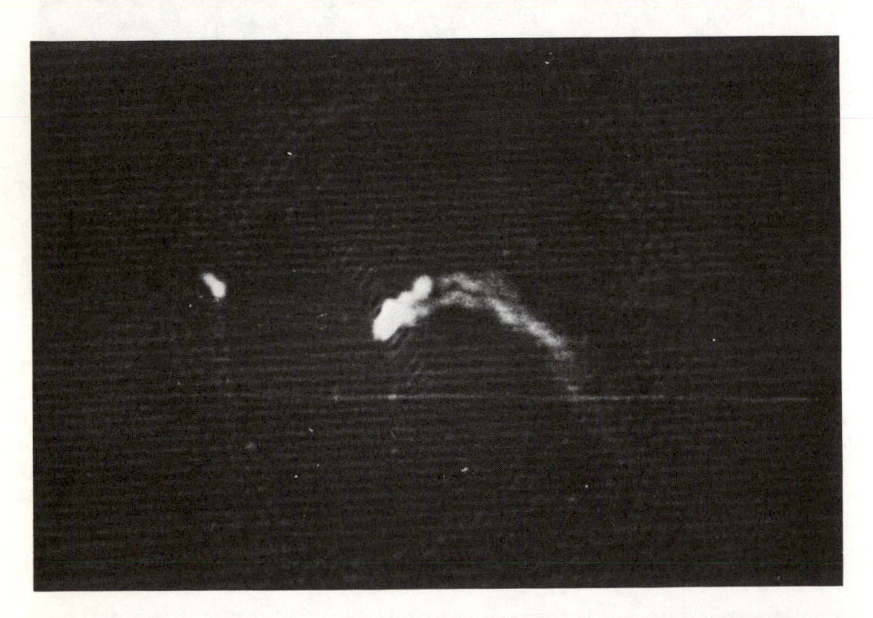

Fig. 64—A radio photograph (radiograph) of the galaxies 3C129 and 3C129.1. The optically visible galaxy is located at the front of the brightest region of this radiograph. We are probably witnessing a trail left by the galaxy moving thrrough the dense intergalactic medium within a cluster of galaxies. The radio tail is about a quarter of a degree long and stretches 400 kiloparsecs through space. (Courtesy Westerbork Radio Observatory.)

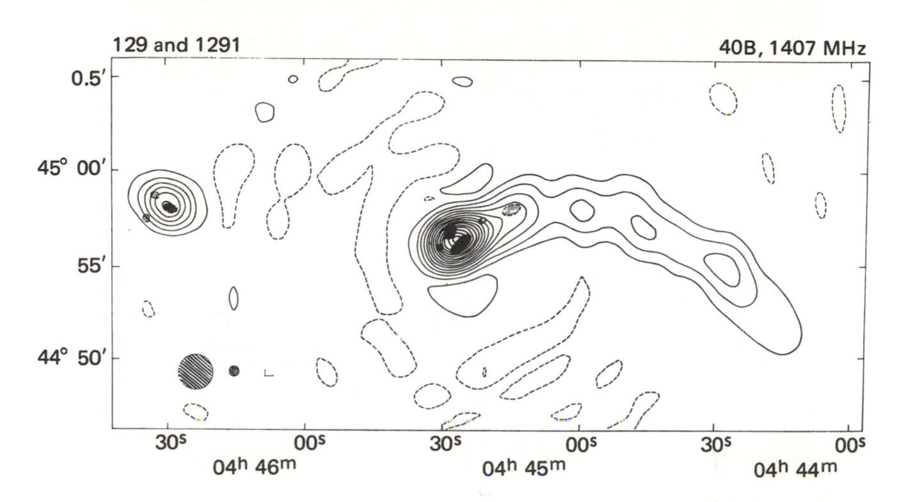

Fig. 65—A contour map of the pair of radio sources 3C129 and 3C129.1 made with the Cambridge one-mile synthesis radio telescope at a wavelength of 75 centimeters. This map has had some areas shaded in; these indicate where the Cambridge radio astronomers found a small-scale structure on a similar map made at a wavelength of 21 centimeters. This contour map may be compared with the radiograph for these two galaxies made at Westerbork. (Courtesy Mullard Radio Observatory.)

Manchester at Jodrell Bank. The telescope, when it reaches the construction stage, will be located in an isolated valley in Wales and will be used as a single instrument as well as in conjunction with the telescopes at Jodrell Bank.

Plans have been drawn up at the NRAO for one of the most beautiful instruments that is ever likely to be built, although at the time of writing the funding situation was not very hopeful. This is the 210-foot telescope designed by Sebastian von Hoerner. This telescope will operate down to a wavelength of 3 millimeters; thus it would be by far the best short-wavelength dish in the world. To have a telescope capable of operating at this wavelength the surface has to be very accurate, and the normal distortions produced by gravity or wind forces have to be taken care of by very sophisticated design techniques. The design incorporates the fact that a structure can be built in which distortions can be controlled in such a way that they cancel themselves out or, at worst, distort the parabolic shape of the dish into another parabola, slightly different, but still perfect. This principle, known as homology, has been used to a limited extent on other telescopes, but van Hoerner's design is the most revolutionary yet. One of the problems with building a very large telescope capable of operating at such short wavelengths is that it needs to be pointed with the same accuracy as an optical telescope and also that one needs to be able to adjust the surface plates to the nearest fraction of a millimeter if the dish is to be used at a

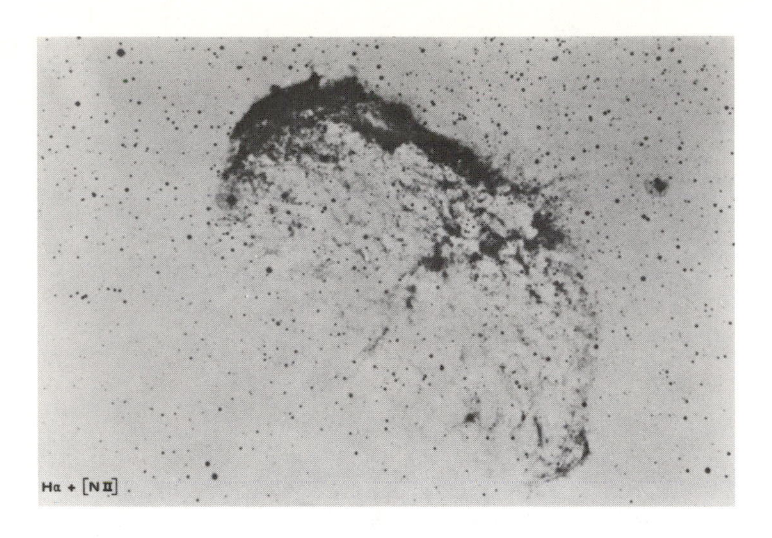

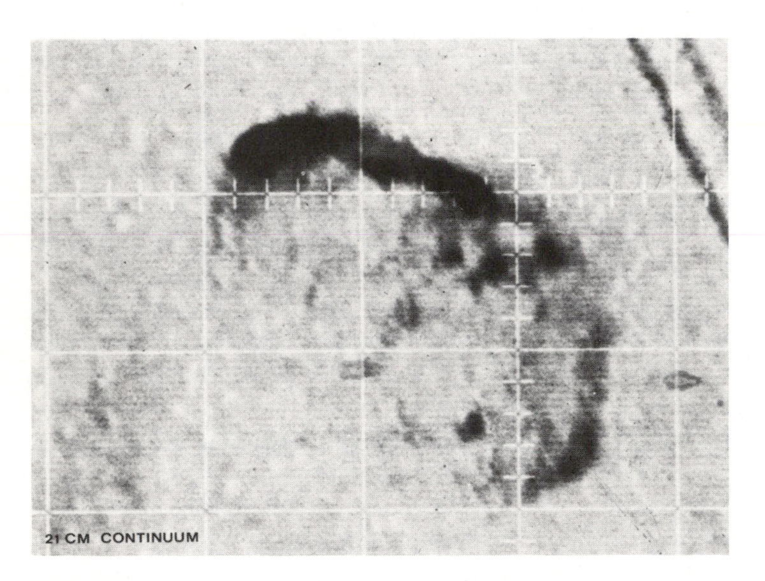

Fig. 66—An optical photograph (negative: upper picture) and a radiograph of the nebula NGC 6888 located in Cassiopeia at a distance of 1.7 kiloparsecs. The size of the pictures is 12 by 18 minutes of arc. (Courtesy Westerbork Radio Observatory.)

wavelength of a few millimeters. To do this, one requires a very accurate, but convenient, distance measuring device for measuring the location of the surface plates with respect to the focus. A device for doing this has already been invented and built at the NRAO. Indeed all the technical aspects concerned with the construction, control, and adjustment of the telescope seem to be ironed out. Only the 10 million dollars required is not available yet!

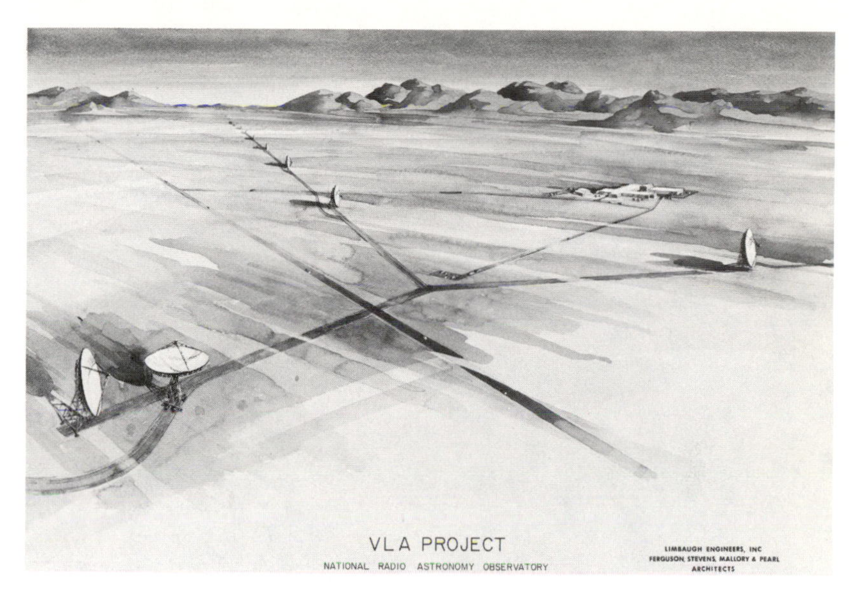

Fig. 67—An artist's impression of the Very Large Array now under construction in New Mexico. (Courtesy NRAO.)

Project Cyclops

As a purely engineering and systems project exercise, a group of engineers as well as several radio astronomers, met in Stanford, California, in 1971 to discuss the design problems that would be faced when building 1000 dishes and locating them in a tightly packed array so as to collect the weakest possible signals from space. They reached the remarkable conclusion that for about two billion dollars a radio telescope system sensitive enough to pick up TV and FM leakage from any planet similar to the Earth, but located up to 100 light years away, could be built. By leakage we mean those signals that naturally head into space away from the transmitters. A telescope system like this would allow us to tune into the TV programs of other civilizations—and who knows what fascinating soap operas they are televising to their captive audiences!

The other strong point about having such a radio telescope is that deep space probes could be tracked, carrying only very weak, and consequently light-weight, transmitters to the farthest reaches of the solar system. This means that for a given expenditure of so many dollars we would not throw most of it into space in the form of heavy payloads, but would have much of it left at our disposal in the form of a tremendously powerful radio telescope for probing deeper into space and for searching for other civilizations. The time may not be far off when man will seriously consider this possibility.

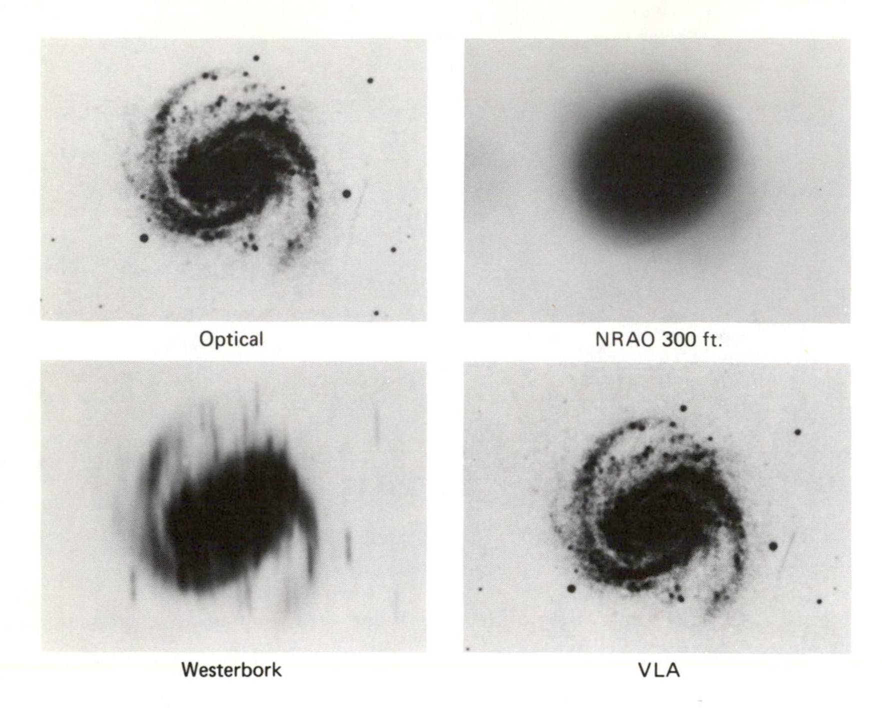

Fig. 68—An illustration of the differences in resolving capability of some radio telescopes. In the upper left is an optical photograph of a galaxy. In the upper right picture we see what the 300-foot telescope would detect if it mapped that galaxy. The bottom two maps show what the Westerbork telescope would do and what the planned Very Large Array (VLA) would be able to resolve. (Courtesy NRAO.)

CHAPTER 18

HOW A RADIO ASTRONOMER MAKES HIS OBSERVATIONS AND STUDIES THE DATA

Preparation

This chapter describes the typical way experiments are performed at the NRAO, which is a facility open to all qualified astronomers. The modus operandi at other observatories, especially those run less democratically, might differ in different degrees from this outline. However, in most large observatories, which have expensive telescopes, the process is very similar to the one described here.

First, it is necessary to have an idea for an experiment, and the evolution of such ideas is not always clear. Then the astronomer must prove that the idea had a rational birth and grew out of some well thought-out piece of research. Once he has the idea for an experiment, he must determine just which telescope and equipment will be required, how much observing time will be needed, and generally how he plans to make the observations. For example, he must decide how the actual observations might be made, and which parts of the sky, or which radio sources, should be observed. These points must be written out and submitted to the director of the observatory in the form of an observing request. At the NRAO this observing request is sent to several outside persons who anonymously referee the request; that is, they determine if the request is scientifically feasible, well thought-out, and able to be done in the manner specified. These referees then rate the request on some scale of values referring to how important it is to the particular field of research.

An astronomer may run into problems at this point because the referees, being human, might not always agree with the astronomer submitting the request. Referees have been known to reject experiments as being too outlandish or ridiculous to warrant being given any time; then a few years later someone else, somewhere else, does the same thing and makes an important discovery. To avoid this as much as possible the NRAO uses several referees. The requests are also sent to several on-staff members, who check that the equipment being requested is indeed suitable, or even exists; and in some cases, special auxiliary equipment might have to be constructed.

Once every few months the telescopes are scheduled for the next several months, and the astronomers who have been scheduled to use the equipment are notified so that they can make their travel plans to visit the NRAO. University faculty members must also usually rearrange classes.

Observation

Before he arrives at the telescopes to begin observing, the astronomer will have prepared a detailed program of observations to be made, made sure of the form the data-taking at the telescope, and determined which computer programs are available to facilitate the basic data handling. For more specialized experiments (and most experiments are special cases!) the astronomer will have to write additional programs to process the data after it goes through the basic reduction procedures available on a large computer. One of the other aspects of observing programs is the preparation of lists of coordinates for pointing the telescope and a list of frequencies to which the receiver is to be tuned, especially if the experiment involves the observation of some sort of spectral lines. Such calculations must be made because the Earth is always spinning on its own axis and traveling in orbit around the Sun, and the Doppler effect seen to affect a spectral line emitted from a distant cloud of gas is modulated by the motion of the Earth. This must be taken into account before and after the observations are made.

Arriving at the telescope, the astronomer, now referred to as an observer, will find that the desired equipment has been installed by the engineers responsible for the construction and subsequent maintenance of the particular receiver and associated equipment. If the equipment is behaving badly, the astronomer might be able to do something about it, provided he is experienced enough; otherwise the engineer is called to the telescope.

The control of the telescopes at the NRAO is left to professionally trained telescope operators who know all about the inner workings of the telescope and its control system. The operators are the ones who will keep a log of the observations being made and make sure that observations get onto the data recording devices. These are usually magnetic tape recorders that are controlled by a small on-line computer, which might also control the telescope motion.

The observer will usually keep close track of the way the experiment is progressing, but will leave a set of observing instructions for the operator when he himself is unable to be there.

Every day the tapes from each of the telescopes at Green Bank are sent to the main laboratory of the NRAO in Charlottesville, Virginia. There the permanent scientific staff have offices; there also are the computer and its staff of programmers. The day's tapes are given priority in being run through the computer, and someone in Charlottesville might call the observer if the data appears seriously inaccurate. These first passes through the computer are usually very standard and the output will make the 120-mile journey back the next day. The tapes, both the one from the telescope and the one generated with the programs run on that first day, will stay in Charlottesville. So it continues until the end of a particular observing run. The observer might have used the telescope for a full 24-hour day or he might have shared the time with other observers wanting the telescope during different parts of the day.

After completion of the observing program, the observer takes copies of the log sheets, the output of the programs, and perhaps the tapes themselves back to his home institution for further data processing. Or he might choose to stay in the Charlottesville lab to work on his data.

Publication

Further work usually involves presenting the observations in such a way that the observer can see the effects he wanted to. For example, he might make a map of a given part of the sky in order to study the structure of a radio source or he might compare the map of an object taken with two different polarizations in order to measure the polarization of the source being studied. Alternatively, if he had been trying to detect line radiation from a new molecule, he might require that all the data that was collected over the period of a week be added in order to reduce its random noise, thus enhancing the signal with respect to the noise, as was described before.

If he is successful in finding what he looked for, the observer might ponder on the physical significance of his discovery and then write a descriptive paper about his observations. This paper is then sent to an astronomical journal. The editor will first send it to a competent referee, who will evaluate the paper and decide if it needs revision before being published or if it should be rejected.

A paper is usually accepted—usually subject to revision. The astronomer will modify it and send it to the editor, and from there it is sent to the printer. Galley proofs are then sent to the astronomer for corrections, and the paper appears in an issue of the journal a short time later. There it either will be read and often referred to by subsequent workers in the field, a fate reserved for fewer and fewer papers, or it will be relegated to the limbo that is the inside of a thick volume of the year's papers in any scientific discipline. The author receives reprints of the paper, which he may send to colleagues or to libraries.

The astronomer, now fascinated by his work and new discoveries, will perhaps have further bright ideas about very worthwhile experiments and will submit further observing requests, which will bring him more satisfaction and sometimes progress in the general field of astronomical research.

There is no doubt that the thrill of making a fundamentally new discovery is enormous. However, at first usually little is gained from studying the data at the telescope and it is only after several days or months of data processing on the larger off-line computers that results are realized.

There are relatively few experiments performed in recent years that have generated the enormous excitement associated with an on-the-spot discovery. One of these was the discovery of radio signals from interstellar formaldehyde, which was seen in absorption in the direction of a dust cloud when it should have been detected in emission. This was because the dust cloud was so cold that it apparently absorbed the 3 degree background from the universe (this is the result of an inverse-maser effect in the molecules themselves). Another such discovery was the detection of a strong magnetic field in the direction of the Orion Nebula in an experiment that I did in 1968. After 7 years of trying the experiment, with no success, I had succeeded in May 1968, but found the signals only after several months of work in Charlottesville. I returned to Green Bank to make further measurements in December, this time with equipment and computing facilities at the telescope, which allowed me to examine the data as it was being received from space. The first few hours I checked the very weak magnetic field effect previously found in the direction of Cassiopeia A. I then turned the telescope toward Orion, in whose direction several very dense hydrogen clouds were located. My attempt to find a magnetic field in these clouds was immediately successful, the signal being clearly visible after only 30 minutes of observation. After 7 years of unfruitful work on various telescopes, with no means of knowing what had been seen while observing, because the signals were too weak, the detection of the effect in the direction of Orion was mind-blowing. For several days I had to make additional measurements to be sure that the effect was real; subtle effects occurring in space or in the antenna system could have given a false answer.

CHAPTER 19

IS ANYONE ELSE OUT THERE?

Life in the Universe

The subject of this chapter could easily be expanded into a book, and several authors have done so already. The question is basically: Are we the only inhabited planet in the universe, and if so, why? Or are we but one of countless inhabited planets throughout the universe, and if so, where are the others? If there are others, how can we ever communicate with them, and are we likely to pick up signals from them in the near future? Since I shall discuss these topics from my own point of view and draw relatively little from the opinions of others, this chapter is not a summary of "official" opinion.

My approach will also be to examine to what extent radio astronomical observations throw light on the possible existence of life elsewhere and what light might be thrown in the not too distant future.

Interstellar Molecules and the Formation of Planets

The most obvious clue now being gathered that bears on the problem is the discovery of so many species of molecules in space. Many of these molecules appear to be confined to well-defined clouds, clouds that may well be the precursors to star formation. This in itself is not too significant, but what is significant is that the molecules in question are mostly organic; that is, they are carbon-based. All life on Earth depends on the chemistry of organic molecules. Indeed, the molecules found in the extreme conditions of outer space are just those needed to synthesize the simplest molecules for the first stages of molecule building on Earth. Those essential molecules are the proteins and the amino acids. To construct these, chemists require such substances as hydrogen cyanide, ammonia, formaldehyde, and water. Molecules such as cyanoacetylene and methyl alcohol are also important in the laboratory synthesis experiments, and all these molecules are found in clouds in interstellar space.

One commonly pictures life on Earth beginning as the result of chemical evolution in the so-called primeval ooze or as the result of the combination of important elements in the atmosphere occurring from lightning discharges. Alternatively, the action of ultra-violet light from the Sun enables various elements to combine into the basic molecules required for the life-forming molecules to emerge. These are the more complex amino acids, proteins, and subsequently, DNA molecules.

Many of the basic two-, three-, four-, and even up to seven-atom molecules required in the stages of molecule building before life starts already exist in space before the planets form. The question now is whether these molecules can survive the heat generated by the formation of the stars. In particular the ultra-violet light generated by the stars can destroy many of these molecules in open space. However, most of the molecular clouds so far observed by radio astronomers are associated with dense dust clouds (see

Fig. 69—Two globules that are very dense clouds of interstellar dust thought to be the precursors to star formation and in which molecules are found. Photographed by B. J. Bok with the 90-inch telescope at Steward Observatory, Tucson, Arizona. (Courtesy B. J. Bok.)

Figure 69), which no doubt act as effective shields from the radiation of stars.

The important question is why these molecules are in space at all. How were they formed? Did they form in the clouds in which we now find them, or were they created in a protostar that often lies within such clouds, and were they then ejected into surrounding interstellar space without being destroyed? If the molecules are formed in the clouds in space before the stars are formed and if some of them can survive in the outer parts of planetary systems that are subsequently formed, then one can imagine that they would be available on the surfaces of these planets soon after their birth. They would therefore be immediately available for further molecule building to form amino acids, etc. This would save us having to envisage the "primeval ooze" as being the chemical plant on Earth.

At present astrochemists doubt that molecules could survive the star and planet formation processes. They think that molecule building would probably have to start all over again once the planet is born. Nevertheless, the fact that organic chemistry has occurred in the alien environment of interstellar space, as well as in the Earth's atmosphere and seas, does argue very strongly that such chemical processes are very common throughout our galaxy and probably the universe. This suggests that the type of molecules and subsequently lifeforms that abound on Earth might abound equally well on other planets throughout the galaxy.

What evidence is there for other planetary systems in our galaxy? Many stars are known to be binary pairs, and most current theories on star formation suggest that stars form as pairs or that a single star forms together with planets. It is also possible that binary stars have planets in complicated orbits about them. Therefore, there are likely to be billions of planets in the galaxy.

There is direct evidence for at least two other planetary systems associated with nearby stars. Both Barnard's star and Epsilon Eridani in the way they are observed to move across the sky show indications of the existence of one or more companions of Jupiter-like mass. Recently the proper motions of a number of other stars have been measured with sufficient accuracy to suggest that they too have planets in orbit about them. Measurements made on these stars during the next several years will help astronomers deduce more accurately the precise sizes and numbers of planets orbiting these stars.

There are several indirect pieces of evidence concerning the likelihood of planets formed during the evolution of most stars. For example, the youngest, brightest stars often seem to be surrounded by a large amount of dust, whereas stars in later stages of evolution do not seem to have this dust around them. It is possible that we no longer see the dust because planets have been formed around the older stars. Another well-known phenomenon is that the youngest stars are spinning very fast, whereas stars like the Sun spin very slowly. It has been noted that if all the mass of the solar system were dropped into the Sun, then it too would be spinning as fast as the young stars. Is it therefore possible that as the stars age they slow down because some of their matter has been ejected and has subsequently formed planets?

These rather loose comments might serve to illustrate that planetary systems might be the rule rather than the exception. After all, for those few stars for which it has been possible to make sufficiently accurate observations we find that all have planetary mass objects in orbit about them. It has also been argued that based on the average nature of our star and the average nature of our galaxy, there is little reason to suppose that our planetary system is anything but average. Some recent theories on planetary formation have even come up with the conclusion that a star like the Sun would always have a planetary system similar to ours. In other words, we would always find an Earth-like planet at a distance from the parent star like our Sun-Earth distance and there would always be a Jupiter-like planet at the equivalent distance from the stars.

Combine these speculations with the fact that molecules abound in some of the proto-stellar clouds, then one finds that there must be billions of planets with conditions not very dissimilar to those on Earth. And we know that the probability of life developing on Earth is unity, that is, it is certain. After all, it has happened! If the probability approaches anywhere near this on these other planets, then life must abound throughout the Milky Way and the universe. But where is it? Can we hope to contact these other civilizations?

Searching for Extraterrestrial Life

Is it worthwhile to search for the existence of life elsewhere in the universe? What good will it do us if we find it? Are we the ones that must search or should we transmit signals toward the planets orbiting other stars? Could we ever communicate significantly with a civilization many light years away? Would it be dangerous? These are some of the questions we can ask, and everyone can try to answer them for himself. I will give some of my answers. First of all, I believe that any contact that is made is likely to be accidental, in that we will pick up signals not specifically intended for us. We might tune in on communications signals to their spaceships or we might pick up their radar or TV. However, it has also occurred to many astronomers that the worst enemy of interstellar communication is cable TV, because there is no leakage into space from such a system. Even if there were cable TV on Mars, we would not pick up anything on Earth! So an accidental discovery is possible if they have evolved in ways not too different from ours. If their planets are going to have a chemistry similar to Earth's, and are going to be relatively similar to the Earth in size and distance from their star, then I envisage lifeforms basically similar to those on this planet. Needless to say, different species may have dominated other planets.

An important point not to be overlooked in asking if we are able to pick up signals from other civilizations is whether any other civilization has reached our level of technological development, and if it has, whether it considered it worthwhile to continue along that line. For example, the popular notion that we are likely to contact a civilization very much further advanced than ours might be a fallacy, because who is to say that further advancement means infinitely more technological development. Perhaps these other planets will also have to face the pollution and population problems we now face and perhaps they will solve them by developing a less technological society, which will stop using up its natural resources at a suicidal level, and hence shy away from building more and more sophisticated machines for their comfort and amusement.

Because of this possibility, it is my belief that we are likely to make contact only with those at about our stage of evolution. This enormously reduces the probability that we will ever make contact, because we must rule out most of the civilizations within a finite distance, since they would have either passed far beyond our acquisitive stage of development or they would be very far behind us. My thesis is that although civilizations, or at least life, is universal, there are likely to be few cases at which life is at a level where communication with us is possible.

To return to the idea of searching for these lifeforms at similar stages of development: How can we do it? It is clear that radio signals, and less likely light waves, are going to play a part; but what sort of radio signals will these be?

A suggestion made many years ago was that the most obvious wavelength at which to transmit signals to inform other civilizations of one's presence was the natural wavelength of 21 centimeters at which hydrogen atoms transmit. That was the most obvious choice during the 1950s, when this was the only wavelength at which radio astronomers had found a spectral line. Now we know of the 18-centimeter spectral lines of OH, the 1.3-centimeter lines of water and ammonia, and a host of other naturally occurring spectral lines. As I mentioned before, the OH lines might be a better bet for contacting other civilizations because the OH molecule emits signals at four closely spaced wavelengths in very definite intensity ratios. Another civilization need only change the intensities of transmitted signals at these four wavelengths to some obviously strange values and perhaps vary them with time a little, so anyone receiving them would realize they were artificially generated. Of course, this is precisely what was observed by radio astronomers, but very adequate theories have now been put forward to explain the anomalous OH lines as the result of a natural phenomenon—in this case, the result of maser amplification in the cloud of OH itself. Such a masering effect has also been observed to occur in the radiation from several other species of molecule, notably interstellar water.

Alternatively, the other civilization might transmit a steady stream of pulses, a sort of carrier wave on which time variations might carry the information. Or perhaps variations of the pulses with wavelength (frequency modulation) might be the form of communication. Pulsars vary this way, and as was mentioned before, for a while they were suspected of being artificially generated. Now they have reasonable explanations, although the explanations proposed are sometimes just as iincredible as an artificial origin!

These examples of how man's ingenuity can find a physical explanation for nearly any unlikely observed event makes one wonder how we would ever recognize an artificially generated signal if we were to receive one. Again, my point of view is that the discovery of signals from other civilizations will be accidental and that it might be years after such a signal has been received before we realize that we are in contact with another civilization. I cannot stress enough that it is the accidental tuning in to signals from other planets not meant for us that will prove that other lifeforms exist. And what then? How do we acknowledge or inform them that we are receiving something? We shall return to this later.

During the latter part of 1971, I undertook an experiment to try to pick up radio signals possibly transmitted toward Earth from three nearby stars, Epsilon Eridani, Tau Ceti, and 61 Cygni. I used the 300-foot telescope of the NRAO at Green Bank, to track each of these sources for about 4 minutes a day for six weeks, with the receiver tuned to the hydrogen line wavelength. The experiment utilized some brief periods of time in an otherwise full observing program to map clouds of neutral hydrogen in the Milky Way. I discovered no variable signals at all and no signals that could not be attributed to normal hydrogen emission from interstellar clouds. A similar experiment was performed in 1960 by Frank Drake, who observed for a few weeks with very much less sensitive equipment and also found nothing. The difficulty with making such observations is not the problems concerned with the possibilities of life elsewhere, or the probability of success, but with the skepticism of the astronomical community in general toward experiments like this. Many argue that the telescopes should be used for reputable scientific experiments and not be wasted on wild goose chases. Will such an attitude ever be altered? I have seen it suggested that the most likely way to make time available for such experiments is to build a telescope especially for such projects. The only way that that could be done, without upsetting the establishment in the scientific community, would be to have the telescope privately funded by some donor who had several tens of millions of dollars to spare—although a smaller amount of money would enable limited searches to be made. The question is one of suitably instrumenting a large telescope and, once that was done, of maintaining the equipment at peak efficiency with competent engineers and trained radio astronomers. I find this particular approach to the question of a systematic search for other civilizations the most logical in a country like the United States. In Russia astronomers are much more likely to consider these matters seriously, in that they are actually starting a search.

Should We Transmit Instead?

Why should we be the ones that have to search for radio signals from space? Why not take the initiative and transmit the signals that will inform others of our existence? This would require a very long-term experiment by a highly motivated group. There would be little chance of success in 20 or more years. It is unlikely that we will transmit in the foreseeable future, which brings us to the question of why some other civilization should be more likely to attempt it any more than we are. Again, it seems more reasonable that we will pick up signals from elsewhere which were not meant for us at all.

Can We Receive Their Signals Now?

If another civilization had been pointing a 300-foot telescope located on a planet 10 light years distant directly at the 300-foot telescope at the NRAO, which I was using for my experiment, I would have picked up his signal if he had been transmitting only 1 megawatt of power! There is absolutely no problem, technically that is, in picking up radio signals from other civilizations located even many tens of light years from Earth. In fact, a telescope system designed to detect the leakage of TV and FM signals from other civilizations has been designed and in principle could be constructed at any time.

Will mankind become sufficiently excited by the possibility of learning about life elsewhere in the near future to support experiments to search for it? It would be a tremendously exciting quest for any country or group of countries to undertake. It would be a project in which private funding could possibly help, especially if the attitude of the general scientific community or the general public turned away from expenditure on such problems. I think we should do it.

Index

Absorption in the atmosphere, 16
absorption of radio waves, 16
absorption of radio waves in space, 52
Aldebaran, 130
Algol, 130
analog–to–digital converter, 8
Andromeda galaxy, 63, 94
angular size of quasars, 105
Antares, 130
antenna temperature, 11
antennas, radio, 5, 7
  TV, 4
  Wurzburg, 24
aperture synthesis, 149
Arp, H., 109
astrochemisty, 83
astronomical unit, 44

Baade, W., 89, 124
bandwagon effect, 83
Barnard's star, 163
baseline, of data collected, 9
baseline, of interferometer, 105
Bell, J., 115
Bell Telephone Laboratories, 21, 140
Beta Lyrae, 131
Betelgeuse, 130
Big Bang; *see* cosmology, 81, 135
birth of radio astronomy, 21
black body emission; *see* emission processes
Bolton, T., 132
Brahe, Tycho, 55
Burke, B., 41
bursts from Jupiter; *see* Jupiter bursts
bursts from the Sun; *see* solar bursts

Cassiopeia A, 56, 75, 89
Civilizations; *see* extraterrestrial civilizations
clusters of galaxies, 96
Cocke, W. J., 125
colliding galaxies, 92
communication, interstellar, 164
computors in radio astronomy, 9
continental drift, 109
corona Borealis, 36
corona, solar, 32
cosmic rays, 51
cosmic rays, radio echoes from, 26
cosmic ray showers, 27
cosmology, 135, 140
cosmology, big bang, 81, 135
  expanding universe, 135
  primeval fireball, 140
  static universe, 139
  steady state, 136
  three degree background, 140
Crab nebula, 41, 47, 54, 59, 89, 123, 129
Crab nebula; polarization of light, 54
  pulsar, 121
Crab nebula, pulsar distance, 127
  X-ray source, 55
Cyclops project, 155
Cygnus A radio source, 89, 96
Cygni, 61, 166

diameter of quasars, 108
direction finding, 17
Disney, M. J., 125
dispersion in interstellar medium, 121, 123, 126
dispersion measure, 127
Doppler effect, 61, 63, 76, 102, 128, 158

dust clouds, 71
dust cloud in Taurus, 81
Dwingeloo radio observatory, 24

Einstein, Albert, 103, 141
electromagnetic radiation, 1, 68
electromagnetic spectrum, 1, 16
electron spin, 59
emission of spectral lines, 49
emission processes, black body, 40
  non-thermal, 40, 47
  plasma oscillations, 35, 48
  synchrotron, 47, 51, 54, 103
  thermal, 40, 47
Epsilon Eridani, 163, 166
expanding universe; *see* cosmology
explosions in galaxies, 93, 97
explosions in radio sources, 103
extra-terrestrial life, 117, 122, 126, 155, 161, 165
eye, human, 2, 18

Faraday, Michael, 72
Faraday effect, 127
Faraday rotation, 72
flare stars, 129
flares, solar, 34
flux unit, definition (Table 2), 90, 91
formaldehyde, 79
Franklin, K., 41

galactic rotation, 63
Galaxy, the; *see* Milky Way
galaxy, M87; *see* Virgo A
  Andromeda, 63, 94
  radio; *see* radio sources
  Seyfert, 96
galaxies, clusters of, 97
  colliding, 92
  explosions in, 93, 97
Gaussian emission lines, 62
Giacobinids, 27
Green Bank, *see* National Radio Astronomy Observatory
greenhouse on Venus, 42
Gregory, P., 132
guest stars, 54

Hazard, C., 101
Hewish, A., 115
Hey, J. S., 25
high velocity clouds, 64
Hjellming, R., 130
Hoerner, S. von, 103, 109
homology telescope, 153
Hubble, Edwin, 93, 135
Hulst, H. van de, 25, 58
hydrogen; *see* interstellar neutral hydrogen
hydroxyl; *see* interstellar molecules

interference, 20, 41, 105, 116, 122
interference fringes, 88
interferometers, 29, 88, 98, 130, 148
interferometers, aperture synthesis, 149
  baseline of, 89, 108
  intercontinental, 106
  lobewidth of, 105
  long baseline, 104
  radio link, 105
  resolving power of, 89, 105
  short baseline, 148
interstellar communication, 164
interstellar magnetic fields, 51, 68, 71, 127, 160
interstellar magnetic field strength, 72
interstellar medium, dispersion in, 121
  pulsars as probes of, 126
interstellar molecules, 74, 78, 161
interstellar molecules, hydroxyl, 74
  masers, 77
  mysterium, 75, 83
interstellar neutral hydrogen, 25, 58
interstellar neutral hydrogen photograph, 67
interstellar neutral hydrogen temperature, 61
interstellar polarization, 71
interstellar space, 58
ionization in the atmosphere, 26
ionosphere, 16

Jansky, Karl, 21, 50
jets in radio sources, 94
Jodrell Bank, 24, 26, 128, 155
Jodrell Bank, as satellite tracking station, 28
Jodrell Bank, story of, 27
Jupiter bursts, 41
Jupiter radio signals, 40
Jupiter temperature, 41

Kellermann, K., 139
Kepler, Johannes, 55
kinetic temperature, 48

life in space, *see* extra-terrestrial life
lightning, 21
little green men, 115
Lovell, Sir B., 25
lunar occultations, 89, 101

magnetic fields in flares, 35
magnetic fields in space, *see* interstellar magnetic fields
mapping, 11
Mars, radio signals from, 43
maser amplification, 77
maser, inverse, 81
Massachusetts Institute of Technology, 75
measurements in radio astronomy, 9
meteor showers, 27
Mercury, phase effect, 44
Mercury, rotation, 44
microwave spectroscopy, 75
Milky Way, 21, 50, 111
Milky Way, mapping of, 51
  radio signals from, 50
  rotation of, 50
  size of, 50, 63
  structure of, 63
Minkowski, R., 124
Mira, 130
Mullard Radio Astronomy Observatory, 24
mysterium, 75

National Bureau of Standards, 78
National Radio Astronomy Observatory, 24, 29, 80, 107, 121, 130, 142, 157
National Science Foundation, 29
Naval Research Laboratories, 25
nebula, Crab; *see* Crab nebula
  Cassiopeia, *see* Cassiopeia A
  Orion, *see* Orion nebula
neutral hydrogen, *see* interstellar neutral hydrogen
neutron stars, 118
noise, 8, 36
noise from space, 23
noise storms, 36
non-thermal emission, *see* emission proceses
novae, 130
Nuffield Radio Astronomy Laboratories, *see* Jodrell Bank

Oort, J. H., 66
Orion nebula, 89, 129, 160

Palomar atlas, 86
Penzias, A., 140
phase effect, 44
phase velocities, 72
photosphere, 32
Planck's equation, 138
planetary radar, 44
planets, formation of, 161
planets, radio signals from, 39
plasma oscillations, *see* emission processes
polarization, 14, 68
polarization, of light from Crab nebula, 54
  of starlight, 71
  of sunglasses, 69
  rotation, 72
  types of, 70
positions of radio source, *see* radio source positions
primeval fireball, *see* cosmology
protected bands in radio astronomy, 19
pulsars, 115, 118, 125
pulsars, as probes of the interstellar medium, 126
  Crab; *see* Crab nebula
  dispersion of signals, 127
  list, 119
  magnetic fields, 125
  names, 121
  periods, 118

quasar 3C48, 99
  3C273, 100, 142
  3C279, 142
  journey to a, 111
quasars, 98, 129
quasars, angular sizes of, 105
  diameters of, 105
  distances to, 101
  locating, 100

quasars (*cont.*)
positions of, *see* lunar occultations
stellar collisions in, 110

radar astronomy, 44
radio antennas, *see* antennas, 4
radio astronomy, birth, 21
computors in, 9
protected bands, 19
radio blackouts, 36
radiographs, 149
radio observatories, Arecibo, 46
California Institute of Technology, 106, 142
Cambridge, England, 24, 86, 116
Cornell, 106
Crimea, USSR, 107
Dwingeloo, 24
Goldstone, 106
Jodrell Bank, *see* Jodrell Bank
Massachusetts Institute of Technology, 75, 106
Mullar Radio Astronomy Observatory, *see* Cambridge
National Radio Astronomy Observatory, *see* NRAO
Naval Research Laboratories, 25
Nuffield Radio Astronomy Observatory, *see* Jodrel Bank
Naval Research Laboratories, 25
Parkes, 86, 101
Pentiction, 128
Westerbork, 149
radio signals from, Mars, 42
Milky Way, *see* Milky Way
planets, 39
Saturn, 42
space, 7
Sun, 24, 32, 36
radio source, 8, 23, 56, 129
radio source, counts, 136
Cassiopeia A, *see* Cassiopeia A,
Cygnus A, *see* Cygnus A
Centaurus A, 97
definition, 85
diameters, 108
double, 94, 97, 101, 137
explosions, 103
jets, 94
M82, 97
M87, *see* Virgo A
Orion A, *see* Orion nebula
position, 85, 89
quasistellar, *see* quasar
scintillations, 116
spectra, 11
radio source, variability, 13, 103
Virgo A, *see* Virgo A
radio stars, 23, 85, 129
radio telescopes, 3, 5, 18, 143
radio telescopes as thermometers, 39
radio telescopes at the NRAO, 29
radio telescopes, not shiny, 15
Jupiter, 40
rat race, 26
Reber, Grote, 21, 23, 50
recording equipment, 6
redshift, 62, 102, 135, 137
reflecting telescope, 4
Reifenstein, E., 121
relativity, *see* Einstein, Albert
resolving power, 18, 105
response pattern of antenna, 19
Rigel, 130
rotation measure, 127
Russian spacecraft, 27
Ryle, Sir Martin, 24

Schmidt, M., 102
scintillations, 116
secrecy syndrome, 83
Seyfert, galaxies, 96
Shklovsky, I. S., 47
sidereal day, 21
Smith, F. G., 89
solar, bursts, 35
corona, 32, 142
flares, 34
infra-red radiation, 34
solar, radio signals, 24, 32
storms, 34, 36
spectral line observations, 159
spectral lines, 49, 62, 74
spectral lines, *see* interstellar molecules
spectral lines, *see* interstellar neutral hydrogen
spectral line list, 84
spectral line redshifts, 137
spectrum, 11
spectrum, electromagnetic, 1

spectrum, windows in, 16
spurs, 52
Sputnik, 27
Sramek, R., 142
Staelin, D., 121
starlight, twinkling, 116
starquakes, 126
stars, 130
stars, Barnard's star, 163
  Epsilon Eridani, 163, 166
  flare, 129
  neutron, 118
  novae, 130
  radio, *see* radio stars
  61 Cygni, 166
  Tau Ceti, 166
steady state, *see* cosmology
stellar collisions, 110
sun, quiet, 32
sunspots, 24
supernovae, 54
synchrotron emission, *see* emission processes

Tau Ceti, 166
Taurus dust cloud, 81
telescope, homology, 153
  radio, *see* radio telescopes
  reflecting, 5
telescope operators, 158
temperatures, kinetic, 48
thermal emission, *see* emission proceses
three degree background, *see* cosmology

Uhuru satellite, 133
universe, *see* cosmology

variability of radio sources, 13
Venus, greenhouse, 42
  phase effect, 44
  radio signals, 42
very large array, 132, 150
Virgo, A, 93, 96

Wade, C., 130
Westerbork, *see* radio observatories
white dwarfs, 115
Wild, J. P., 36
Wilson, R., 140

X-ray sources, 131
X-ray sources in, Crab nebula, 55
  Cygnus, 131, 133
  Scorpius, 131